Little Rice

Smartphones, Xiaomi, and
the Chinese Dream

COLUMBIA GLOBAL REPORTS
NEW YORK

Little Rice
Smartphones, Xiaomi, and the Chinese Dream

Clay Shirky

Little Rice: Smartphones, Xiaomi, and the Chinese Dream
Copyright © 2015 by Clay Shirky

Published by Columbia Global Reports
91 Claremont Avenue, Suite 515, New York, NY 10027
http://globalreports.columbia.edu/
facebook.com/columbiaglobalreports
@columbiaGR

Library of Congress Control Number: 2015946812
ISBN: 978-0-9909763-2-5

Book design by Strick&Williams
Map design by Jeffrey L. Ward
Author photograph credit: Barbara Alper

Printed in the United States of America

CONTENTS

Smartphones

A couple of years ago, while I was doing some work at NYU's Shanghai campus, I got lost on the subway. As a New Yorker, it takes a lot to make me feel like a country mouse, but at triple the population of my hometown, Shanghai does it. Even though the Shanghai subway system is amazingly well-provisioned with directions in English, I got out at the wrong stop. I didn't figure this out right away, because the subway exited into a mall, just like at my stop, and Shanghai has so many malls—36 million square feet of retail space will be built this year—it can be hard to tell one from another.

Walking in a daze through a vast collection of hallways and shops, I did the very thing people who build confusing malls wanted me to do: I slowed down and started looking around, whereupon I noticed a booth selling mobile phones, a thing I happened to need at the time. I saw a particularly nice one, all black, rounded sides, quite stylish, whose logo read Mi3. I decided that a Mi3 would be as good a phone as any, so the vendor and I did that curious pointing and gesturing thing people do when transacting with no common language except money, and ten minutes later, I had my phone.

There is not much a middle-aged guy can do to seem *au courant* to eighteen-year-olds, but that phone did it. For the next several days on campus, whenever I needed to do anything on my phone, one of the Chinese students would ask, "Where did you get that?" Not, "What kind of phone is that?"—they all recognized it. The Mi3 was a huge hit for Xiaomi, the startup that made it, selling faster than the company could produce them. I had managed to get my hands on a phone so popular, the company couldn't always keep up with demand, making me briefly the envy of teenagers (not a familiar feeling, before or since).

Xiaomi (pronounced like the "show-" in shower, plus "me") is the thing that many people in the West don't think exists: a company that can create products that aren't only made in China, but designed in China, and beautifully so. For decades, the rap on Chinese manufacturing has been, "Oh sure, they can make lots of copies cheaply, but they can't design new products." Over the forty years that China has been open for business, the country's manufacturers have mastered increasingly complex sourcing and assembly for increasingly complex products, especially electronics. (The iPhone box may say "Designed in California," but it is built in Shenzhen.)

For anyone watching this rising mastery of quality, the question has become, "When will Chinese design rival its counterparts in the rest of the world?" Owning a Mi3 made it clear that, at least for electronics, the answer was "2013." It was of

10   high quality and moderately priced—more expensive than most
smartphones, but at 2000 yuan (about $330), it was cheaper
than a similar Samsung, at around $400, and much cheaper
than an iPhone, at over $500—but those virtues are virtues of
purchasing and assembly. The Mi3 is also beautiful.

All smartphones are a slab of black glass with three or four
buttons on the case, so phone design tends toward rearranging
these minimal elements. The Mi3's minimalism was to make a
thin phone seem even thinner by making the screen look as if it
ran from one edge of the phone to the other. On many of Xiaomi's
early phones, and most strikingly on the Mi3, the edges of the
phone case curve away so sharply from the screen that the eye
discounts them as part of the same surface. This was a trick, of
course—you can't make a cheap phone if the case doesn't stick
out past the screen—but it was a good trick, and more importantly, it was a trick that meant that people inside Xiaomi were
thinking, very carefully, about what a good phone would look like.

The mobile phone is a member of a small class of human
inventions, a tool so essential it has become all but invisible,
and life without it unimaginable. The common desiderata of
the world's adult population (and of most of our children), the
mobile phone is the site of a steadily increasing amount of the
world's communication, from selfies to contract negotiations.
Jan Chipchase, an ethnographer who has studied the use of
mobile phones worldwide, points out that there are only three
universally personal items that someone will carry with them
no matter where they live. The first two are money and keys;
the third is the mobile phone, making it the first new invention
added to that short list in three thousand years.

Since launching in the late 1970s in Japan, mobile phones have become the fastest-spreading piece of consumer hardware ever, faster than automobiles or fixed-line phones or even televisions. Because individual wires do not have to be run to individual houses, and because the cost of the handset is shared with the user, mobile phones are far cheaper to deploy than fixed lines, and thus often connect populations that never had connection before. American teenagers have long insisted that they couldn't live without their phones, but this phrase has real meaning in the developing world, where the kind of information you get from a phone can have a profound effect on the quality of life: Fishermen in Kenya use phones to figure out where they can sell their catch, parents in India use it to locate doctors in other towns, and so on. Mobile phones may be a big improvement over fixed phones, but they are a gigantic improvement over no phones at all.

This dramatic change in what is awkwardly called teledensity is now almost universal. The number of mobile phone users crossed 4.5 billion last year, and because of dual accounts, there are now more mobile subscriptions in the world than there are people. In sub-Saharan Africa, mobile phone penetration is around 66 percent in 2014—two subscriptions for every three people in a region where the phone network extends much further than the electrical grid, leaving it to small-business people to sell phone-charging services using car batteries. Penetration for the world's heavily indebted poor countries—HIPC to the United Nations, the poorest of the poor to you and me—is only just behind at 58 percent, or three mobile subscriptions for every five people, in countries with barely functioning

12 economies. Meanwhile the countries with the lowest penetration are there not because of lousy economics, but repressive politics. North Korea, Myanmar, Eritrea, and Cuba are the only populous countries with less than 25 percent adoption. Absent direct repression by the state, the citizens of the world are adopting mobile phones at a torrid pace.

All these phones have to be made someplace, and that place is China (as with most things that get made). Of all the things made in China, some are culturally specific enough to resist export; the global market for busts of Chairman Mao isn't all that much larger than the Chinese market. Others are universal; there is no country-specific version of a 5 mm screw or a Hello Kitty pencil. In between the Chinese-only products and the universal ones, though, are products that could go either way, products made in and for the Chinese market, but which might become global exports. Mobile phones straddle this divide.

Most mobile phones are made *in* China, of course, but some are made *for* China. There are the cheap knockoffs, part of China's tradition of *shanzhai* manufacturing. *Shanzhai* refers to mountain villages that make their own goods, and carries the sense of inexpensive and expedient production, including a less than robust concern for patents and trademarks. Some of these goods are simply cheap, barely functional phones, but some are copies meant to look (at least at a distance) like their more expensive inspirations. At the open-air electronics market on Baoshan Street in my adoptive town of Shanghai, these copies take dozens of incarnations. Samsung is a favorite recipient of this flattery, with knockoffs bearing logos like San Song and Svmsmvg. These phones tend toward national markets; neither the Svmsmvg nor the San Song will spread much outside China. Other lures for selling cheap phones are used elsewhere; the Kenyans were offered an Obama-branded phone a few years ago, but it's not a strategy for all markets.

Then there are the phones designed for East Asian sensibilities. The same region that brought us the selfie stick also brought us Oppo, a company whose phone's principal selling points include a high-quality camera and custom software that automatically airbrushes photos with faces in them. The ad campaigns emphasize a particularly performative form of femininity, since, in a nice touch, the software makes a guess about the gender of the subject—everyone gets smoother skin, but only the ladies get their lips reddened. Despite successful rollouts in Thailand and Korea, Oppo has not made much of a dent in markets outside East Asia. Their U.S. launch was a bust.

Mobile phones, in other words, have mostly been just another Chinese export—the cheap products for poorer markets are thrown together at minimum cost, while expensive products for the increasingly global group of well-heeled customers are designed elsewhere, whether in Seoul or San Jose, the pattern that led Apple to add the phrase "Designed in California" to its packaging in the first place. This pattern of "designers elsewhere, manufacturers here" has been the norm since the British turned southern China into their workshop in the 1800s, but it is starting to shift. A number of Chinese companies are moving to do everything at home, working to create mobile phones where "Designed in China" means quality, not just *shanzhai*. (This is the same path famously trod by Sony in the 1970s, when its founder was determined to retire "Made in

14   Japan" as an insult.) The most successful of these new design-oriented companies, and therefore one of the most important mobile phone manufacturers in the world right now, is Xiaomi. Xiaomi is the first Chinese phone manufacturer to compete, globally and successfully, not just on price but on innovation in design and service.

The full name of the company is Xiaomi Tech but everybody just calls it Xiaomi, not unlike everyone saying Apple back when they were still Apple Computer. (The parallels between Xiaomi and Apple are one of the constant themes of discussion around the company, and a sore point for the founder.) The literal translation of *xiao mi* is "little rice," the Chinese word for millet. The theme of grain exists in their product branding—their current line of cheaper phones is called the Hong Mi, meaning "red rice," and millet itself is a cheap grain, harkening back to the Chinese revolutionary slogan "millet and rifle," indicating an army willing to live on a subsistence diet.

Xiaomi, founded in Beijing in 2010 by Lei Jun, a computer scientist and charismatic serial entrepreneur now in his mid-forties, has accomplished a lot in half a decade. Even just looking at its sales figures, the superlatives pile up. In its short life, it has gone from a startup focused on making a new mobile phone interface to beating Samsung as the number one phone vendor in the largest market in the world in 2014. Its products are so popular in China that it has become the third largest ecommerce firm there just selling its own products, after the general marketplace sites Alibaba and JD.com, and ahead of Amazon.cn.

As the company has adopted increasingly international
aims, the name became something of a liability—few English
speakers are used to pronouncing words that begin with an
"x." So it changed its public face to emphasize Mi as its brand,
including buying Mi.com last year for $3.6 million, one of the
most expensive domain name acquisitions in Chinese history.
(Richard Liu, an early investor, winces while recalling the purchase, as he'd been an early supporter of the Xiaomi name, one
of the few missteps the company made on its way to a global
market.) The company says Mi refers to "me," to mobile interface, and to *Mission Impossible*. Enthusiastic Mi users are called
Mi Fans in English, and Mi Fen ("rice noodles") in Mandarin.
Not since Humbert Humbert was chasing Quilty has there been
this much wordplay in so few words.

On November 11 (11/11, known as Singles Day, and by 2014
the largest online sales day in any country in the world), Taobao,
a service like eBay, but considerably larger, and run by Alibaba,
sold almost 1.9 million phones, of which nearly 1.2 million were
Xiaomis. That same year, five of every eight Android phones
activated in China were Xiaomis. In early 2015, the company
launched a "phablet" (the tech industry's ugly term for a cross
between a phone and a tablet) called the Mi Note, priced at 2,300
yuan ($442), cheaper than Samsung's S5, at over $600. The first
day the Note was offered, it sold out in three minutes. Xiaomi
raised $41 million in their first round of funding in 2010. As
2014 closed, it was not yet five years old and valued at $45 bil-
lion. It is, by several metrics, the most valuable startup ever.

The rise of Xiaomi in half a decade is more than a business
story, because mobile phones are special. A mobile phone offers

16   the kind of freedom and connectedness that autocratic countries have historically been terrified of, and China is no exception. The Chinese government has spent the last twenty years
building the largest and most pervasive system of surveillance
and censorship in the world. Yet China's fortune and future
clearly lie with the opening up it has conducted for the last forty
years. This requires the government to allow local entrepreneurs to have connections with the outside world for import
and export, and with them, the freedom to experiment.

China sends an increasing number of its citizens abroad,
most notably dispatching many of its brightest students away
for education in the world's democracies, and especially the
U.S. And yet the current government is clearly convinced that
it cannot afford the same kind of openness to new ideas among
the general population, so the economic opening up is being
accompanied by the sort of broad and deep media crackdown
not seen in China since the late 1980s. To give you an idea of the
scalpel-like removal of the free play of ideas from public conversation, the government agency in charge of media has banned
both time travel and the use of puns in certain kinds of writing,
as both forms allow for alternate meanings that might threaten
government narratives.

There is no particularly communist approach to handling
mobile phones or the internet, mostly because their spread
postdates communism as a going concern. Out of two dozen
communist countries a generation ago, plus another dozen
absorbed as Soviet republics, only five are left. Four share borders in East Asia: China, with Vietnam and Laos to its south and
North Korea to its east. (Lonely Cuba, about to be swallowed

whole by Club Med, is the extra-Asian holdout.) Copying China, both Laos and Vietnam have long since dropped opposition to public markets and private business. Only North Korea is sticking with the disaster that is collectivized agriculture. (It dismantled communism through nepotism rather than markets.) "Communist" is a merely historical label for how those five governments came to power, providing little predictive utility for current policies or behaviors. All four Asian countries have preserved some of *The Communist Manifesto*'s dictum to centralize communications in the hands of the state, at least for old media; newspapers, radio, and TV remain heavily controlled. (To this day, parts of Laos get "community radio" via loudspeakers mounted on poles.) However, their collective reaction to the internet varies, to say the least.

North Korea has implemented as close to an outright ban as any country in the world. (Myanmar, its nearest regional competitor on that score, recently started opening up, *à la* China.) Meanwhile, Vietnam, with a more tourist-driven economy, allows significantly fewer restrictions in areas where Vietnamese and outsiders interact. (The free Wi-Fi in the Ho Chi Minh airport exhorts visitors to "Like us on Facebook!") China, the Middle Kingdom, is in the middle on this as well, constantly titrating what they allow and deny in order to remain open for business and closed to criticism.

A mobile phone is a kind of lens that makes the importance and contradictions of modern China easier to see. Every year, *The Economist* calculates the price of a single item—a Big Mac hamburger—across every country where it is sold, as a way to compare the relative strength or weakness of those countries' currencies. The rationale behind the resulting "Big Mac index" is that this simple item is really a bundle of products and services, reflecting the cost of flour and beef, of wages and taxes, and even rent, electricity, and security. What looks like a product on the outside is really the endpoint of a complex chain of events that includes the cost of acquisition, assembly, and sales.

If all that goes into a hamburger, imagine the lines of influence that converge in a mobile phone. Hardware and software, operating systems and applications, a "quality vs. price" tradeoff for a hundred different components, an international distribution chain of astonishing complexity. All this just to produce a device that then has to be marketed, shipped, and sold in a dozen or even a hundred different countries, accompanied by the increasing importance of apps for everything from dating to journalism; and there are conflicting demands for freedom (by users) and control (by governments). As a bundle of products and services masquerading as a simple object, a mobile phone is a thousand times more complex than a Big Mac. Hamburgers illustrate something about global economics; mobile phones do all that and illustrate something about geopolitics as well.

Internet

Public communications is—always—political. Mobile phones are prized because they deliver certain kinds of capabilities to their users: freedom to communicate, to access information, and increasingly to buy and sell, all enabled by a small screen and a hidden antenna. And yet many of the world's governments have committed themselves to monitoring the increasing flood of information, some of it from international sources but most of it from their own citizens. The U.S. version of this, as we now know, is ubiquitous surveillance of nearly everyone the National Security Agency can collect data on, but this is mainly a difference in execution—that sort of surveillance is the goal of almost every government everywhere. (One of the many commercial disasters that befell BlackBerry in recent years was India's demand that BlackBerry Messenger security be unlocked for the government, thus eviscerating one of the company's key services.) The Chinese go further, accompanying widespread surveillance with adaptive and near-real-time censorship.

20   This is an example of what China scholars call post-socialist China, where the rationale for Marxism has been dropped but the institutions of state control have remained. Post-socialist China traffics in the trappings of an ideological state, periodically lauding Mao Zedong Thought without trying to put the Great Man's ideas to any practical use. Absent an ideology other than unbroken Chinese Communist Party rule, the government's commitments to politics are considerably more vivid when discussing what the Chinese state must not do.

The Arab Spring was particularly terrifying to them in this regard, as activists were so reliant on electronic tools to coordinate anger among citizens. Those uprisings looked like plausible threats, especially since the principal site of political action became the main square in the capital city, a pattern that linked November 7 Square in Tunis and Tahrir Square in Cairo with Tiananmen Square in Beijing. (Tianamen was the site of three potentially destabilizing uprisings in the 1970s and 1980s, the last ending in a massacre of hundreds of protesters by the People's Liberation Army on June 4, 1989.) In the aftermath of the Arab Spring in 2011, Beijing circulated a list of the Five Nos, five things that the Chinese government would not allow public discussion of: No system in which multiple parties govern in turn; No diversification of guiding ideologies; No separation of the "three powers" or creation of a bicameral system; No federalization; and No privatization. (That last one is a corker, since China has been privatizing like mad since the 1970s; in communist parlance, it means "No abandoning state control of the commanding heights of the economy," which commanding heights are, by definition, whatever the current government decides not to privatize.)

The list of Five Nos was a precursor to 2014's release of the Seven Don't Mentions, instructing the media not to discuss a similar list of political options. And to prevent the elaborate system of censorship of political subjects itself from becoming a target of popular ire, discussion of why discussion is not allowed is also not allowed. (The Sixth No is No discussing the Five Nos.) Studies of censored topics on the Chinese internet consistently find some of the tightest censorship around discussion of censorship itself.

The operative international rule, normal for centuries now, is that nations don't interfere in each other's internal affairs. This bargain, hashed out in 1648 as a series of peace treaties signed in Westphalia, Germany, gradually became the consensus view almost everywhere. The twin notions of national sovereignty and non-interference allowed wildly different systems of government to exist side by side—monarchies next to democracies next to autarkies—while still allowing trade to pass between them. As the Westphalian idea spread, the world was slowly carved up into a set of territorial claims that ultimately absorbed the whole of its populated surface. Under this system, all humans are citizens, and their passage through the world is restricted by rules governing both emigration and immigration. Every person should have a nation; people without regions are regarded with suspicion (viz the Roma), regions under no clear national ruler are regarded with anxiety (Somalia), and regions claimed by more than one nation are treated as standing crises (as with the fight over the Spratly Islands).

22   But the current round of globalization is much more about goods than about people. Money and trade have always moved more freely than people, and especially so ever since the world's economies liberalized. The Westphalian system rests on two assumptions that have weakened over the centuries: First is the idea that what happens inside a country can be kept separate from its effect on the world, and second, the idea that economics and politics are different spheres.

In the twentieth century, and especially after the Second World War, these fictions didn't put too much strain on the system. It was at least possible to pretend that national politics were separate from international politics, and that we could have global economics but mostly local politics. (That was the founding ideology of the European Union.) Since the opening up of the 1970s, China has emphasized this distinction between economics and politics. When paramount leader Deng Xiaoping originally designated the southern city of Shenzhen as a Special Zone, he was pressured by conservatives in Beijing to change the label to Special Economic Zone, to emphasize that political experimentation would not be tolerated. During the Cold War, communist countries rejected Westphalian logic, since revolutionaries were supposed to be building an international movement. But as that expansionist model ended in the 1980s, China has become a great advocate of Westphalianism. It advocated economic liberalization of the world economy, so that all production that could be done cheaply flowed to it, while insisting on the political sanctity of national borders, to create barriers to access to local Chinese markets for foreign goods and foreign ideas.

You can see this at work most dramatically with China's internet policy. Many observers, including me, believed that of the two categories on offer under the Westphalian system—the free movement of money, and the restricted movement of people—that information would behave more like money. (This was an easy belief to hold, since money is increasingly made of information.) China, remarkably, has managed to create an alternate path, building a country where information moves like people, in highly identified and constrained ways, with the government always reserving the right to refuse entry from elsewhere, along with the ability to apprehend rogue information locally. They have achieved this in part through deep technical competence, in part through consistent national investment, and in part because schemes that sound daft in an American context—hire an army of people to flood social media with positive comments about the government—are achievable given the availability of cheap labor. Here again the scale staggers; bathing in decades of sweat after the 1989 Tiananmen massacre, the Chinese spend more on internal security than on their military, and they have the largest standing army in the world.

For a long time—up until this decade, in fact—the Chinese didn't have much trouble keeping computers under control. Citizens didn't own many, telecom providers were state-owned, and the cost of hardware and bandwidth meant that many young men (the principal source of any government's internal worries) had to use the internet in cybercafés (big business in China until a couple of years ago), which simplified surveillance. As long as computers were pricy and rare, government oversight was relatively manageable. This sort of surveillance only worked so long

24   as most citizens didn't have computers, which in turn meant that it only worked so long as mobile phones were just phones. That too remained true up through the end of the last decade. Up to about 2009, phones only came with two communicative features—calling and texting. Smartphones change that calculation.

A smartphone is as different from a standard-issue Nokia 1100 as a computer is from a typewriter. While the keyboard and screen of a computer provide a familiar analog to the keyboard and paper of the typewriter, the ability of the computer to take on new functions with the addition of new software makes the computer a different kind of thing.

A Nokia was a phone that only had the functions Nokia put into it. (There was always talk of allowing third-party software on Nokia phones, a sort of primitive app store, but the carriers usually disabled those features.) A smartphone, though, is a computer. It's not *like* a computer, it *is* a computer, and an increasingly capable one at that. And the fundamental feature of computers is that they are unpredictable, not in that "my printer stopped working" way, but in a "my old machine can do new things with new software" way. (The two are related—no one has ever made a computer whose flexibility only produces good surprises, and no such device is on the horizon.)

This flexibility makes the smartphone special worldwide. No device goes further to deliver what are effectively superpowers to ordinary individuals than a smartphone. (The average mobile phone famously has more processing power than the spacecraft that went to the moon.) China is now both the largest producer and consumer of these networked computers we still call phones. The government simultaneously recognizes that you can't be a modern economy if your citizens don't have networked computers in their pockets and that you can't keep political coordination amongst your people at bay unless you can keep coordination at bay, full stop. All of this makes smartphones a special class of problem for the Chinese government, because phones help people replace planning with coordination. After people get mobile phones, they make fewer definite plans, preferring "call when you're here" to "meet me at six." This unpredictability appears in the political realm as well, in events both large (the sudden massing of protesters in the Arab Spring, coordinated digitally) and small (public sentiment in China on issues like corruption and pollution are increasingly hashed out online).

Because online conversation in China is increasingly fast and informal, the usual modes of censorship and surveillance are no longer enough to keep control of public opinion, and the government is expanding its online propaganda efforts. The people who flood online conversations with pro-Beijing sentiment are called the 50 Cent Party, on the (appealing but false) idea that they are paid half a yuan for every post. (The point has a bit of wordplay, as the Chinese phrase is the "5 Mao Army," where a *mao* is both the name for a dime and of the father of the country.) But even this is not enough, so the government is now recruiting members of the Communist Youth League to be part of a volunteer online propaganda group, each of whom would agree to participate in at least three "sunshine comment" campaigns a year. The participants will be volunteers in the sense that they are unpaid, but the Communist Party is not leaving it to chance—there are quotas for both regions (Beijing

26 is expected to recruit 250,000 members) and for particular colleges and universities (Sun Yat-sen University is on the hook for 9,000). Alongside the hundreds of thousands in the 5 Mao Army, the government is hoping to recruit one-fifth of the Communist Youth League, which would amount to 18 million volunteers, and would be a propaganda operation the likes of which the world has never seen.

President Xi Jinping has been far more aggressive than his predecessor in pushing for control of media, especially the internet. In November 2013, a year after he came to power, he announced the founding of a new government department, the Central Leading Group for Internet Security and Informatization, with himself as its leader. He also centralized various models of governmental control under a single group—the Cyberspace Agency of China (CAC). As Bill Bishop, a longtime China watcher, puts it, the party believes that if China can't transform the internet, the internet will transform China. More recently, the People's Liberation Army has taken to declaring the internet a new front for the struggle against China's enemies. An editorial in the *PLA Daily* claims that criticism of the government on social media brings about "the tearing apart of social consensus, the tearing apart of the relationship between the Party and the masses, and between the Army and the people, in a bid to overthrow China through the Internet."

These were not one-time adjustments. The Xi government is determined to increase political checks and controls on technology firms. Foreign companies like HP and Cisco are subjected to greater scrutiny, while local firms are finding themselves subjected to new rules. In July of 2015, the government

introduced the final version of a long-debated national security
law, which included provisions for a national security review
and "oversight management" of a variety of industries, including "internet information technology produces and services."

The long period in which digital technology firms were
treated as normal (which, in China, is to say export-friendly) is
ending; the party has decided that the need to ensure that internet technologies are "secure and controllable" outweighs the
business requirements of globalization. For companies that sell
basic infrastructure, this dictum provides an edge when competing with Western firms such as Cisco and Alcatel. For firms
like Xiaomi, though, that aspire to sell to individual consumers,
the assurance that the party considers their services secure and
controllable will not offer much of a selling point outside China.

CAC's guiding principle is what's known as "cybersovereignty," the idea that the internet should have borders and
controls for information just as it has customs and passports
for people. As has been widely noted for decades, this works
against the basic architecture of the internet, making something like cybersovereignty incredibly hard to accomplish. Any
country that could successfully do such a thing would have
to have a curious combination of deep international isolation
(no Arab country can escape pan-Arabic media like Al Jazeera,
because the language is not limited to one regime), significant
local technical talent (Turkey does not have enough engineers
to build credible local alternatives to most of what it wants to
block), and an investment climate that would provide nascent
national internet firms with the capital needed to get started
(however badly Sudan might want local internet firms, no one

28 has the capital to fund such a thing). The one country that has this combination of talent, entrepreneurial zeal, robust private investment, and censorious attitude toward media is China. They have succeeded in making a regime of dynamic censorship and propaganda that is far and away the most successful the world has ever seen.

This has been a long time coming. The Chinese idea of a non-globalized internet had its first realization in the late 1990s as a set of filters for the web, partly automated and partly run by human oversight, and designed to block the flow of information in from the outside world. This system was called the Golden Shield, but was so quickly and universally referred to as the Great Firewall of China that even some Chinese abbreviate it GFW. This first version of the GFW was aimed at keeping information from the rest of the world out. At this stage, there was a small number of professional outlets outside China producing material on a relatively predictable schedule: Wikipedia, the *New York Times*, *The Economist*, and so on. Though the Great Firewall has implemented a general blockade of selected Western internet news services since then, its most consequential function has been to block social media from the outside, especially Facebook and Twitter. The real threat to China is no longer popular access to information but popular access to coordination. Major events like the Sichuan earthquake in 2008 to a deadly high-speed train crash in Wenzhou in 2011 were first reported by citizens using their mobile phones and Chinese social networks like QQ, Renren, Sina Weibo, and WeChat. No amount of external filtering can seal off this threat; the calls are coming from inside the building.

The change in the operation of the GFW represents a shift in Beijing's main worry, from controlling information from the outside world to preventing synchronization of opinion and, even more threatening, coordination of physical assembly by citizens. And of course, the locus of the threat has also shifted from personal computers to mobile phones. The blockade, in place since 2006, simultaneously prevents Chinese citizens from having access to communications tools that the government can't censor or shut down, and provides a competitive space for homegrown social media to flourish, with services like Renren, Weibo, and WeChat acquiring hundreds of millions of users as the Chinese internet market has exploded in the last decade. The question of whether the blockade of Facebook is economic or political is pointless—it is both. Even were Facebook to accede to all of China's political demands, the economic value of having homegrown alternatives is too large to easily reverse.

Any company marketing communications tools inside China necessarily subjects itself to government oversight that is intrusive, detailed, and mercurial. ("To work in China, you need two habits," a friend in Beijing says. "First, you have to be optimistic. Second, you have to be forgetful.") It is no overstatement to say that companies that want to offer communications services both inside and outside China end up having to formulate something like a foreign policy, just to land a contract.

Xiaomi thus sits at the intersection of cheap copies and investment in innovation, local and global demands, market opportunity and political restriction, and freedom and control. The story of Xiaomi is entwined with the spread of global communications.

Xiaomi

For all the elegance of the hardware, Xiaomi is at heart a software firm. Its CEO and all of its original co-founders came from software firms, and the company's first product (and only product for the first year of its existence) was an operating system for a mobile phone. MIUI (short for "Mi User Interface" and pronounced, quite by design, "Me-You-I") is a modified version of Android, itself a modified version of Linux. Android, distributed by Google, now runs the majority of the world's smartphones. (Apple's software for the iPhone is the only popular alternative, and Android phones outsell iPhones three to one.) Android is free for vendors to use—Google, its main developer, gives it away to have a presence on smartphones—and within certain restrictions, such as offering Google software, phone manufacturers are free to customize Android.

For most of Android's short life, this option meant little more than altering the style of the interface. If you moved from a Google Nexus to a Samsung Galaxy, not much changed other than the background and the icons—the apps and the experience were very similar. The few firms dedicating serious design resources to their products tended to concentrate on making the hardware better. Software customization was left to the user, by way of the apps we choose to install. What set MIUI apart was that even before Xiaomi had a phone of its own to sell, they set about making the operating system work better than the competition. For the first year of the company's life, their only users were people interested enough to download a copy of MIUI and install it on their existing phones, replacing whatever flavor of Android had come with the phone. These users were pioneering (and geeky), and Xiaomi paid close attention to what they wanted and how they used the phones.

Some of the improvements in MIUI were plain old-fashioned performance tuning. Xiaomi paid particular attention to making MIUI work well on Samsung phones. (Though the firm often gets compared to Apple, their product line and mid-market ambitions are much closer to Samsung's.) By 2011 MIUI was better looking and more responsive when running on Samsung smartphones than Samsung's own version of Android was, and, critically, didn't drain the battery as quickly—a huge, underappreciated part of the user experience generally. Part of the plan with early users was to get testing and feedback, of course, but another part was to get free publicity for the user experience. (Sample tweets from 2011: "I installed MIUI into my [Samsung] N1, It is like fresh air. I am feeling good." "I just installed MIUI, my phone became much more easy to use immediately. [Very handsome].") This combination of treating users

32   as both sources of feedback and as amateur marketers continues to this day. Hugo Barra, the Google executive in charge of
Android who joined Xiaomi in 2013 to spearhead international
expansion, boasts that the company spends almost nothing
on traditional advertising, preferring to stage launch events
that the press will cover, and helping their users proselytize on
behalf of the company.

It is in this last category that Xiaomi excels. Tony Wei,
Xiaomi's marketing chief, took me on a tour of the company's
offices, and pointed out a cluster of desks where employees
were working on advertising. Those employees weren't graphic
designers or photographers, they were programmers; in keeping
with its founding expertise in software, the company makes its
own tools for interacting with its users, and for helping those
users interact with the world. The company has thought this
through with a thoroughness that almost no other firms take
on; even the most plumbing-like activities can offer a platform for user outreach. After installing a new version of MIUI,
the confirmation screen offers users the ability to send a message on Sina Weibo (China's Twitter, roughly) that they've just
upgraded to a new version.

Of course, none of the marketing would have worked if the
product wasn't good, but in the beginning, it didn't have to be
perfect. It just had to be better than the other Android phones.
As time went on, Xiaomi began adding features that don't appear
on ordinary Android phones: a better note-taking app, their
own music subscription service, their own cloud-based backup
service, a MIUI-specific tool for screening out advertising calls
(China does not have a National Do Not Call Registry, unlike the

U.S.), and, for Asian sensibilities, a highly customizable set of interface themes, since mobile phones are a far more personal item in China than in the U.S.—both phone cases and the look of the home screen are sites of obsessive self-expression.

There was no one "killer feature," no thing that MIUI users could do that other smartphone users couldn't. Instead, early MIUI users got three things that made a difference. The first two were practical: a better experience without upgrading their hardware, and the attentions of a company that was fanatically solicitous of their feedback, seeking out expert users who agreed to provide weekly critiques of MIUI, even before they had shipped the first version, a pattern that continues to this day.

As the user base has grown from the initial hundred recruits to over a hundred million today, Xiaomi began separating its users into two categories—"fever" fans, who are the most eager for new features and the most technically savvy, and "flood" fans, ordinary users who like Xiaomi's products but can't provide detailed feedback. Fever fans are consulted early, given access to products and services while they are still in the initial testing stages. (Some fever fans have proven themselves so valuable that they have been brought on as consultants.) Flood fans are the ones who get the ordinary Friday updates, and post their comments in the Mi forums. Their opinions are generally less technical, but with hundreds of thousands of them active on the forums and discussing the company on social media, their aggregate opinions are useful to Xiaomi, both as research and as marketing. Xiaomi, in turn, is consistently engaged in online community building, creating national "Popcorn" fan events, local Mi Fan communities, and

34   even posting pictures of fake Mi phones they find in electronics markets to warn customers.

MIT economist Eric von Hippel calls this sort of user involvement "lead user innovation." In a number of fields, including cooking, mountain climbing, and industrial robotics, the most intensive users, like the fever fans, often understand the product as well or better than the designers, and the modifications and adaptations made by those users are often good candidates for incorporation into the standard product itself. Bill Joy, a co-founder of Sun Microsystems, once said, "No matter who you are, most of the smart people work for someone else." Lead user innovation is a way to bring some of that outside intelligence to bear. Xiaomi has brought lead user innovation into the mobile phone world; Lei Jun has estimated that something like a third of the features in MIUI come from user requests, and he often credits users with co-designing MIUI.

The third thing Xiaomi created was an ineffable sense of specialness for Xiaomi users, the same sense that companies like Apple, Harley-Davidson, and REI produce in their customers. Doing any two of those things would have given them a product, but not a hugely successful one. The company had to deliver a better experience, be more responsive to its users, *and* reward them with a sense of their perspicacity in adopting MIUI to set themselves apart in a crowded, cost-conscious market.

Even without hardware, Xiaomi began making money by offering new service options to users. The plan was simple: Users would install MIUI to get improvements and customizations, and when those users began paying for downloads and services, it would be the maker of the operating system, not the hardware manufacturer, who would capture those revenues. Taking a cue from Nippon Telegraph and Telephone (NTT) in Japan and SK Telecom in Korea, Xiaomi even invented its own currency, Xiaomi Credits, which users can use to top up their phone, buy music or custom themes, and so on. (One credit equals one yuan, but as in Las Vegas, treating money as "credits" makes users more willing to spend.) More recently, Xiaomi has announced the addition of Mi Finance, a PayPal-like system for MIUI users, another attempt to earn incremental revenues from the people who purchase their hardware.

Xiaomi is thus both beneficiary and driver of the Chinese technology industry's shift from products to services. Lei understood that China was crossing over, that ownership and use of phones would dwarf ownership and use of PCs and laptops, and that the phone would become the locus of new service businesses. Lei had been in the software business since 1992, when he joined Kingsoft, a Hong Kong-based software firm, as an engineer, rising to CEO in just six years. While engaging in many side businesses, he ran the firm until he resigned at the end of 2007, having finally taken it public after four previously unsuccessful attempts. After Kingsoft, he spent two years as an investor and board member of various internet firms, including UCWeb, which made a mobile browser called UC Browser. UCWeb gave Lei a ringside seat to the rise of mobile internet, and by 2010, he was contemplating starting a new firm.

The internet industry in China went from tiny to huge in a hurry, much like the rest of the economy; as a result, many of the heads of the largest firms not only know one another but have business dealings that go back to the earliest appearance

36   of the internet in the late 1990s. In Lei's case, Tencent—China's most important social media company, operators of the ubiquitous WeChat—invested in Kingsoft. Alibaba, the ecommerce powerhouse, which is Amazon, eBay, and PayPal in a single firm, bought UCWeb. Baidu, the Chinese search giant that also runs a Wikipedia-like service, Baike, co-invested with Xiaomi in Cheetah, a spin-off of Kingsoft. And so on. Lei's connectedness and previous track record at building digital businesses meant that he would have been able to raise money for almost anything he set out to do. The trick was figuring out what to do.

One of the early conversations he had was with Richard Liu, a Shanghainese startup funder at Morningside Ventures, and an early investor in UCWeb. Lei Jun knew that China was poised to become, in the parlance of networked businesses, mobile-first, and in particular smartphone-first. (The iPhone had hit in 2007, but the "touch-screen/app store" model that Apple pioneered didn't come to China until later.) Mobile phones are more constrained than PCs; the screens are smaller, the connections tend to be slower, and the users are considerably less patient. A company that focused on improving users' experience on the phone could reap considerable benefits if it could also offer services the users would pay for—downloads, games, storage, and so on. Lei called Liu to talk through these ideas, and talk they did. Liu says he picked up the phone at 9:00 p.m., and he and Lei started hashing out what opportunities the smartphone ecosystem held for China and the world. (Though Xiaomi was largely unknown outside of China until 2014, its ambitions were global from the start.) The call ended at 9:00 a.m. the next day, a through-the-night conversation that helped solidify the idea for the business and its launch strategy. Not incidentally, Lei brought Liu on as an investor in Xiaomi's first round.

Lei's strategy was to build a company that understood that China, still a country where most mobile phones were of the call-and-text-only variety, was about to join the smartphone revolution. Xiaomi was designed to be ready when it happened. The Chinese market for mobile phones had always been tied to simpler versions, and conventional wisdom assumed that the Chinese, poor on average and famously stingy, would continue to prefer simple "monoblock" and flip phones to fancy and expensive smartphones. As late as 2009, Nokia still had three-quarters of the Chinese market, and that company had bet their fortune, already fading in the rich world, on the idea that in poor countries, smartphones would remain a luxury. Tricia Wang, the great ethnographer of Chinese media use, spotted the change when migrant workers she was studying started talking about saving up to buy smartphones. She told Nokia that this change was coming, but they didn't believe that consumer preferences could change that quickly. This turned out to be a catastrophically bad bet.

The median Chinese citizen still doesn't have much disposable income, so demand for high-tech products lags the rest of the world. But when demand does appear, it does so rapidly. (When an economy grows in excess of 10 percent a year, as China's has been doing until recently, the size of the national economy doubles in less than a decade.) In 2010, Nokia still sold a majority of the phones in China. By 2011, its market share had fallen from nearly three-quarters to less than half, with almost all of its lost market share going to smartphones. The increasing

38 disposable income of the Chinese, especially urban and coastal Chinese, coupled with the falling cost of electronics generally, meant that when the Chinese market changed, it changed all of a sudden. By 2012, smartphones had the same three-quarters market share of purchasers that Nokia had held only three years earlier, and most of these phones were using the Android operating system favored by Asian manufacturers. Tens of millions of people traded in Nokias for smartphones during those two years.

However, Xiaomi didn't take advantage of this change by quickly shipping a new phone, because its first product was software. Lei Jun had understood that good smartphones were expensive and the cheap ones were bad. Any company could deliver a phone in between those points—a phone that was a bit better than junk, for a bit cheaper than a Samsung—just by buying better components while shaving its margins a bit. Because that strategy is available to any company, however, it would only provide a momentary competitive advantage. (That margin-compressing option is what makes the smartphone business so brutal, in fact.) If you could build a company that could deliver a nicely designed and easy-to-use phone for much less than its competitors, you could retain a long-term advantage. Easier said than done, of course, but Xiaomi had a plan, whose outlines were visible in MIUI.

To put any complex technology in the hands of mere mortals, most of the complexity has to be hidden. The industry obsession with "ease of use" is a recognition that if a product requires a manual, it is already at a commercial disadvantage. But if a complex product is going to be made to look simple, the simplicity has to be concentrated on the part of the device the user uses. This meeting point—the interface—becomes the essential design challenge. Lei understood that if you want to sell a complex object to the general public, then the interface *is* the product. Turning the small number of possible user interactions—tap, swipe, pinch, drag—into ways of triggering a huge range of interactions—sending a text, buying a movie ticket, crushing some candy—is the thing a user assesses when using a phone.

Though Lei is the leader and public face of the firm (the "#1 Product Manager," as his employees know him), the firm relied on a diversity of talent from the early days. The canonical image of an internet startup is "two people in a garage," but Xiaomi launched with Lei and half a dozen co-founders, including Lin Bin, the former engineering director of Google Global, and Huang Jiangji, former director of development at Microsoft's China Academy of Engineering. (Google and Microsoft in particular have served as a sort of real-life graduate school for many Chinese internet executives.) In its short life, Xiaomi has been expanding its product line, which now includes a TV, a set-top box, an air purifier (a must in urban China), and an exercise-tracking wrist band, but their first piece of hardware, and the thing they are known best for, was their mobile phones.

Xiaomi was founded in the middle of 2010, with international aspirations from the start. The usual advice to entrepreneurs is to take money from only a few investors at first. Lei Jun pursued the opposite strategy, starting out with a surprisingly global group of investors—its funders included Morningside Ventures (Richard Liu's firm) as well as Temasek, a Singaporean

40   government investment firm, and Qiming Venture Partners, a U.S.-China investment firm. Their second round included Qualcomm, a San Diego-based company that makes most of the chips for the world's mobile phones.

In 2011, Xiaomi launched the Mi1, its first phone. Anyone looking at it would have thought the same thing—it was just OK, and the MIUI running it wasn't very good, either. Android, the operating system MIUI is built on, was still playing catch-up with the iPhone. Given the price of the Mi1, however, MIUI didn't have to be better than the iPhone, it just had to be better than other Android phones, especially Samsung's. (As an early Xiaomi user said in 2011, "At that time, the experiences of using Android was quite bad, I would rate it lower than 60 (if the total is 100). Even MIUI still had some distance to reach the quality and fluency of iOS, it was good enough for me to use it.") The Mi1 was hardly a breakthrough product, and indeed, Xiaomi has never had a breakthrough product, in the sense of something dramatically different from what existed before. It was cheap, of course, at 2,000 yuan (about $330) when Samsung's comparable phones were over $400 and Apple didn't yet have much of a presence in China. Price is a short-term advantage, however. Over the long term, what Xiaomi has is what Toyota has, what Wikipedia has, and what Android itself has: continuous improvement. Though it has lately taken to Apple-style press conferences and product launches, Xiaomi's strength has been to hew to the Chinese virtue of "good enough," coupled with a commitment to "…and better next time."

The "good enough" mentality is familiar to anyone who has paid for lodging in China—even in otherwise luxurious business hotels, the faceplates on light switches and outlets may not be lined up correctly, signs that the interior had been finished in a hurry. Housewares in ordinary stores are serviceable but marketed without much attention to product design. In poor and rural areas, rice wine is sold in vats undifferentiated by any sign of quality other than price. (Beware the 5 yuan stuff.) All this *fenqian*-pinching adds up to the pattern that journalist James Fallows dubbed "happy with crappy." What's special about computers—and the small computers we call phones—is that the hardware can be improved with better software. A computer or phone is really just a hunk of possibility; it's the software that makes it work, and better software makes it work better.

After computers were networked, it became possible to pair "good enough" with "and better with an upgrade." Xiaomi's development efforts have always been based on continuous small upgrades to the interfaces and user experience of its devices. The company will occasionally launch big upgrades for MIUI, but most of the really important work happens through constant small adjustments. Xiaomi ships a slightly upgraded version of MIUI every week. (As I write, they have just dropped their 231st upgrade.) Rather than being treated as background, new upgrades are made available on Fridays, called Orange Fridays in the firm and among users, after Xiaomi's corporate color.

For the programmers working on MIUI, their entire schedule moves around the weekly shipping of updates to users— every Wednesday, for example, the features to be included in the upcoming version of MiUI are announced. The new version becomes "feature complete," in the parlance, to give time for internal testing. The appearance of a new version of MIUI is

42   then an opportunity for external testing—on the weekend after an Orange Friday, thousands of user comments, bug reports, and feature requests flood the Xiaomi forums, to be taken into account for a later update of the system.

All businesses claim that their customers matter to them. Few behave as if this is actually true. Customer service is treated as an annoying cost; product features are worked out in the bowels of the firm and then announced to potential buyers only when the product is finished. Xiaomi, by contrast, takes customer feedback very seriously. Every Orange Friday release comes with an option for the users to fill out four questions: How did I feel when I upgraded MIUI? (Ranked as a series of emoticons, from smiley to frowny faces.) What did I like best about the new update? What did I like least about the new update? What feature do I most want? Answers are then collated and presented by the thousands to the engineers on the following Tuesday. The bug fixes get worked on quickly, while the features often take longer.

Orange Fridays are accompanied by slick videos announcing new features and highlighting Mi Fans. These videos appear on Xiaomi's YouTube channel, among other places, even though YouTube is blocked in China. As with many companies, Xiaomi treats censorship as the cost of doing business in the Chinese market, but uses every blocked tool it can in its global expansion. If the Chinese government were to let Google services through the Great Firewall, Xiaomi would update 100 million phones with Google tools on the next Orange Friday.

Online updates also led to one of Xiaomi's signature tactics—the flash sale. By telling users that a new phone is coming out and asking them to register for participation up to a week in advance, Xiaomi gives itself a better view of demand. In January 2015, it sold out of the Mi Note, its priciest phone to that point, in three minutes. In May, Mi Bands, Xiaomi's fitness tracker, sold out in India in seven seconds. Xiaomi was given a Guinness World Record for selling over two million phones in twelve hours on Singles Day, beating its previous record. The sales are not without controversy, however. The company has been dogged by accusations that it offers up fewer units than it has on hand so that it can generate headlines about it running out of stock in record time. In the case of the seven-second Indian sale, only 1,000 Mi Bands were offered.

The arrival of the Mi1 was followed, at the end of 2011, by the announcement that China Unicom, one of the two big mobile phone providers and a state-owned enterprise, would buy a million Mi1s to offer to customers. Xiaomi had, through a combination of user engagement, online sales, and the Unicom announcement, managed to go from offering its first piece of hardware to having massive aggregate demand in the space of three months. The firm never went through the long phase of growth and bargaining with suppliers; its brand became solid gold with suppliers almost immediately. One investor, who asked not to be named, said that Xiaomi is now able to negotiate the million-unit price for electronics even when a new product is in the testing stage (they are working on a watch and a drone at the moment) because the likelihood of Xiaomi launching a product that will be viable for two years straight is something worth sharing the risk on.

Gary Rieschel, another early investor, explained to me that the usual phone manufacturer sells novelty, then slashes prices

44   on older models, discontinuing them quickly to make way for new models. Xiaomi doesn't do that. It has very few models on sale, and saves rapid updates for software, where it's cheaper. "Our cost of sales is less than 3 percent," says Rieschel, the result of keeping everything to one sales channel. Xiaomi then keeps the model around for nearly two years, reducing the price slowly, while the cost of the components fall by as much as 90 percent. The initial margin is tiny, the ending margin is large, and the average margin stays healthy.

This focus on a handful of individual product lines in turn allows the company to stay small. Employees who have been through Xiaomi's hiring process are told that the company's goal is to hire as few people as possible, by concentrating on attracting and retaining talented employees. This sounds like a bit of corporate pablum, but Xiaomi has only about 8,000 employees. Contrast that number with Baidu, China's internet giant, which is headquartered in the same Beijing neighborhood as Xiaomi. Baidu is worth less than twice as much as Xiaomi—$72 billon to $45 billion—but has more than five times as many employees. Even with rising salaries, low-cost labor remains a highly available substitute for all other inputs, so it takes discipline to keep a workforce small in a growing business.

The company began a Mi Fans festival in April of 2012, an annual event that mimics some of the excitement around Apple product launches. At one of the festivals, the company offered plush-doll versions of their bunny mascot, selling 100,000 of them in a few hours. Its next phone, the Mi2, arrived that fall. The Mi2 was an improvement over the Mi1 in the usual ways— a bit faster and lighter, better screen and camera, for the same

2,000 yuan as the Mi1 originally sold for. (The Mi1 remained
on sale at 1,300 yuan.) Lei Jun claimed that they were originally
selling the Mi2 below cost to get users. At launch, the company
sold 50,000 Mi2s in three minutes. The Mi3 came in the fall
of 2013, along with news of the company's plans to branch out,
announcing work on a large, 3D-capable TV. Fall of 2013 also saw
the hiring of Hugo Barra from Google, where he had been vice
president of product for Android, the core of Xiaomi's MIUI, to
oversee the company's international expansion.

The Mi4 broke the autumn pattern, launching in July of
2014. That year also saw Xiaomi's first sales operation outside China, in Singapore. Moving into a new country is a fairly
complicated affair, since it involves setting up local partner-
ships for online selling and providing customer service in
various languages. Tony Wei, the executive in charge of marketing, took me on a tour of Xiaomi's Beijing service center, where giant screens announced the number of inbound
queries they were fielding from each region of China—a map
weighted to the eastern coast, just like the money is—along
with calls from growing international operations, a total of
30,000 calls and 40,000 chats in busy days in early 2015.
Since Singapore, Xiaomi has expanded into markets that have
large but cost-conscious populations. It has started selling phones in India, Malaysia, and the Philippines, and has
announced plans to enter Thailand, Indonesia, Russia, and, further afield, Turkey, Brazil, and Mexico.

Their international rollouts have not been completely
smooth. Xiaomi ran into trouble almost immediately in India,
when a widely circulated report was issued in August of 2014

46   by the security firm F-Secure, announcing that Xiaomi phones were collecting data on Indian users and sending the data back to China. This turned out to be a result of a default configuration that tells the phone to share system reports with Xiaomi using its Cloud Messaging Service, which routes SMS messages through Xiaomi servers to save money. Xiaomi immediately issued a new version of its software that set the defaults for data sharing and messaging to Off, but not everyone got the news. Several months later, the Indian Air Force asked that its personnel not buy Xiaomi phones, even though the default behavior had long since been patched. (The Air Force later rescinded the request.)

Later that month, a user in Hong Kong—a Chinese territory, but with a largely uncensored media—made the same accusations about personal data being sent to servers in Beijing, and the following month, the Singaporean government, home of Xiaomi's first office outside China, also investigated the company for potential privacy violations, after accusations that the company was selling users' numbers to telemarketers. Though Xiaomi reacted quickly to change its software, and announced by that fall that it was setting up servers outside of China to allay some fears, the rise of national sensitivities around data gathering, and the way they are now intersecting with user privacy, indicates how much of a country-by-country slog Xiaomi faces in its expansion.

In early 2015, the company held a press conference in the U.S. to announce that it was selling its fitness band and other accessories, but no phones. It's widely believed that the U.S. and Western Europe present uniquely challenging markets, as their populations are less cost conscious while Xiaomi's competitors are more litigious. Patent-related suits are far more common in the U.S., and Xiaomi is wary of operating in the local market of many firms that would like to drag them to court, including especially Apple.

The Mi3 is a beautiful phone, and doesn't look like much else on the market. It is, as I discovered with my students, a phone you can recognize at a distance. On the other hand, the Mi4 is an embarrassingly slavish knockoff of the iPhone 5S, down to case color and button shape. At a distance of more than a few inches, it invites confusion, featuring the same white façade and similarly rounded corners as the iPhone. "It is theft and it is lazy," Jonathan Ive, Apple's chief design officer, has said of the Mi4's design. "I don't think it is OK at all." Mobile phone industry watchers believe that Apple is not suing Xiaomi in China because they do not want to jeopardize their access to the luxury market there. However, Xiaomi's patent problems in India, and the fact that it sells accessories but not phones in North America, suggest that its expansion is being shaped by the threat of future lawsuits.

The design of the Mi3 is distinctive. The design of the Mi4 is derivative. This seems odd, coming from the same company in the space of a year, but asking which company is Xiaomi *really*—design-centric or derivative—assumes that companies have some essence. Xiaomi is, for the moment, having it both ways. As long as the Mi3 and Mi4 both have a market, why pick between design innovation and copying? Xiaomi avoids the dilemma summed up in that old business chestnut: "Good. Fast. Cheap. Pick two." So long as a customer trusts that the

48   experience is good enough and will get better, Xiaomi doesn't have to pick.

Even when it became a manufacturing firm, Xiaomi has stayed dedicated to the internet service model. Since it sells its phones directly but only online, it's saved from having to hire and train retail staff in every city it wants to sell in, as well as the logistics hassles and the depreciating assets of trying to keep phones in stock in many different locations. There are Xiaomis available in stores, but the retailers buy the phones at a markup. This retail/re-sale model spares Xiaomi organizational headaches, and well as sparing them the usual conflict in price between stores and online sales. Xiaomi's products are always cheaper online, but it is China Unicom and mom-and-pop electronic stores that have to deal with that discrepancy, not Xiaomi.

Xiaomi's Beijing outlet, on the ground floor of a converted wool warehouse in the tech hub of Zhongguancun, is called the Xiaomi Experience Center, and functions as a place to try before you buy online. You cannot leave the store with a phone, TV, or wrist band. The closest Xiaomi gets to offline sales is with launch events, and in keeping with the general strategy of flash sales, these events seem designed to get people to line up around the block as a way of advertising the desirability of Xiaomi products.

The lack of Xiaomi-branded stores hasn't stopped people from trying to fill that gap. In 2014, completely ersatz Xiaomi stores began opening in larger cities around the country. In Shenzhen's Huaqiangbei Road, the largest electronics market in the world, many such stores remain, sometimes more than one per block, with elaborate Xiaomi signs and even fake Xiaomi staff uniforms. This prompted a warning from Lei Jun that none of these stores were run by Xiaomi, who don't retail their own phones. The concern was that instead of buying phones from Xiaomi and re-selling them, as China Mobile does, the knockoff stores would sell fake phones of the San Song/Svmsmvg sort. (The irony of a Chinese electronics firm worrying about copying was lost on no one.) The fear of copying is so pervasive that when I was visiting Beijing to talk with Xiaomi employees, the first thing they wanted to do was inspect my Mi Note to see if I had been sold a fake. It's real, but the inspection took a couple of minutes, indicating how quickly *shanzhai* manufacturers can produce an acceptable simulacrum. One of the things driving thinness as a selling point is that the engineering that goes into making a slim phone is one of the few hard-to-copy skills left in manufacturing.

Another way that Xiaomi differs from the norms of the mobile phone industry has to do with its location. Beijing, where the company is headquartered, is one of two main tech hubs in China. The other is Shenzhen, the southern city most known for electronics manufacturing—it is the location of the largest Foxconn factory, where iPhones and iPads, among other devices, are assembled. The choice is significant—Huawei, the world's largest electronics manufacturer, is headquartered in Shenzhen, as are Meizu and Oppo, two other Chinese mobile phone brands. Shenzhen is the world's electronics hardware bazaar, a hub of expertise and capability for manufacturing phones. Xiaomi, by contrast, is in Beijing because that's where internet firms are, companies that specialize in software, in services, and in user acquisition.

As the mobile phone increasingly becomes the site of media use, in China as everywhere, keeping a rein on what people do with the software on their phones becomes a matter of increasing concern. This is in part because tech firms cluster where tech education clusters. Just as Route 128, the high-tech region of the 1970s and 1980s that housed such companies as Data General and Digital Equipment Corporation, was near MIT, and Silicon Valley stretches between Stanford and Berkeley, Beijing hosts many of China's top tech schools, including Peking University. The other reason, though, is that Beijing is where the government is, and it wants internet services close by. Michael Anti, a Chinese blogger who's often censored by Chinese authorities, notes that many social media services are required to have their servers in Beijing, to simplify surveillance.

Xiaomi's dedication to internet services sometimes gets them in trouble. Its earliest phones shipped with Google's popular apps pre-installed, until the government forced it to stop in 2013. Company representatives responded by putting up a post on their online forums apologizing for having to take the software off: "Hi, MIUIers, You may have noticed that in our recent updates, the Google apps are gone. We're sorry for the inconvenience that has caused, but we have to remove those apps because of China's relevant polices." They went on to teach users how to download an installer from the Xiaomi app store to add Google apps back. Hugo Barra, the face of Xiaomi's international expansion, makes it a point to repeat in public that the company ships all other phones anywhere in the world with Google installed. Indeed, MIUI was built on top of an effort by Google itself. (The search giant requires licensees to install Google apps on Android phones, but it makes an exception for Chinese phones.)

Xiaomi isn't a repudiation of Shenzhen's *shanzhai* culture so much as its next form. The company couldn't do what it does without having direct access to the greatest manufacturing base in the world. Indeed, in one of the many competitive overlaps in the electronics business, many of Xiaomi's phones are assembled by Foxconn who also produce Apple's products. Xiaomi has shown how to take relatively commoditized parts and turn them into a desirable product, through a combination of physical design, continuous improvement of software and services, and something like an aura. Even if the company eventually becomes a moderately prosperous mid-tier supplier—far from a terrible outcome if your home market is China—its ability to create a product that is worth more than the sum of its parts is already showing other startups how it's done. With at least that success secured, the key question for Xiaomi over its second half decade is how well it will be able to move out of China—a huge market but an odd one—and into the rest of the world.

Number One Producer, Number One Consumer

There are a lot of people in China. This fact can seem like a bit of local color, a background association like France and cheese or Thailand and beaches. For people who live here, though, and especially for people who work here, population is a force like gravity, affecting almost everything else. Once you leave the well-trodden but narrow groove that exists to move tourists over the Great Wall and past the Terracotta Warriors, the country runs on its own logic, not because of any deep Chinese "essence"but simply because they've had to adapt to being populous, poor (on average), and increasingly urban.

If you rank the world's countries by population, you get a gentle upward slope from Vatican City to the United States, unbroken by sudden jumps, and then, at the edge of the graph, there is a sudden discontinuity, a quadrupling of country size from the 350 million souls in the U.S. to the massive populations of India and China. The idea of a billion and a third citizens of any one place is such an abstract number it's hard to grasp—to get to the population of the U.S., take China's and subtract a billion. Like most places in the world, China is urbanizing, but, like most comparisons involving China, it is doing so at an intensity that makes ordinary description inadequate. The most populous parts of the country—the big cities of the east and south, plus a couple of inland empires—simply dwarf anything in the U.S. In urban China (an increasingly good approximation of the Chinese population as a whole) this means that any system that works well works under crushing load. China hosts the largest and second largest subway systems in the world, four of the top ten busiest, and is building out mass transit rail in twenty-five cities simultaneously. (Contrast the United States, where we can't even get our high-speed trains to operate at high speed.)

And as quickly as the Chinese are creating new infrastructure, they are still falling behind, because urban growth and density are increasing more quickly. When I started spending time in Shanghai and tried to describe it to friends back home, I'd start with, "Imagine New York, but busy and crowded." Even that, though, understates the case—China has between half a dozen to two dozen cities bigger than the Big Apple, depending on the urban area measured. Three of those—the municipalities of Shanghai, Beijing, and Chongqing—have populations larger than or roughly the same as New York state's. The multicity sprawl in Guangzhou—the Pearl River Delta mega-city just north of Hong Kong—holds 44 million people, larger than the population of California, in just a tenth the area.

When national holidays synchronize movement, it makes for what the Chinese call *ren shan, ren hai*—people mountain,

54   people sea—moving along in a crush of your fellow humans that approaches geographic scale. The long New Year's celebration carries the same "to grandmother's house we go" imperative that Americans associate with Thanksgiving, except with 700 million people trying to get to grandmother's at the same time, making it the largest human migration on the planet.

This enormous demographic denominator can fry Western sensibilities about wealth and poverty. China is now, by some measures, the world's largest economy, and will be so by all measures soon. Divide this wealth by the population, however, and China lags. Its GDP per capita is 20 percent below the world average, putting it on par with Tunisia and the Dominican Republic. Its neighbor South Korea has nearly triple the per-capita annual national income, while Japan, the U.S., and most of northwestern Europe have quadruple. China is rich, but the Chinese are poor.

Or at least poor on average. The fateful moment for the Chinese economy, crippled by central planning and collectivized production, was when Deng Xiaoping, China's long-term leader after Mao's death, announced that the country would pursue "Socialism with Chinese characteristics," which is to say a market economy under an authoritarian technocracy. This was in 1977, as good a year as any for marking the birth of modern China. Deng and his associates undertook a job akin to that of a political bomb squad, laboriously dismantling most of the economic ideology installed by Mao without blowing up political continuity at the same time. That they succeeded is in many ways the single most important political fact of contemporary China.

It was also Deng who said, "It is OK for some people to get rich first." Communist orthodoxy had made general advancement of the rural population their stated goal since the revolution. For Deng to say otherwise was a near-total break with previous economic theory. Since then, and especially in recent years, some people have indeed gotten rich first. China's growth has been accompanied by the creation of such vast, concentrated fortunes that its income inequality now tops that of even the U.S. China's previous two periods of rising inequality were both brief, and the result of collapses in the incomes of the poor, during the Great Leap Forward and the Cultural Revolution. Today is different—the incomes of the poor, even the rural poor, have been rising, but the incomes of the urban rich have been rising far faster.

Income inequality and the urban-rural divide are also mutually reinforcing. The big cities are increasingly where the money is. When WeChat, a Chinese messaging app like WhatsApp, added advertisements to its service, its parent company Tencent charged advertisers 40 yuan (about $6.50) per thousand users an ad reached. Tencent also gave advertisers the opportunity to specify that their ads would only reach people in Shanghai and Beijing, at more than triple the price. Simply being a WeChat user in one of those cities is a good proxy for wealth and splashy spending habits.

This disparity creates huge migrant classes in China, where men leave their villages or towns to work on construction projects in the big cities, and women leave to work as domestic help or factory workers, but they do so without being able to access a range of official government services—including, critically, the ability to enroll their children in city schools. Whole families have a hard time moving, because China's internal passport system, the *hukou*, ties benefits (including schooling) to your birthplace. China was recently riveted and appalled by the suicide of four children in Guizhou province, who were living alone after their parents left to find work. Just as America outsourced its working class to China, China has created a worker class that can't rely on state services either, using the same sort of displacement. Beijing recently relaxed the *hukou* system, but it is still enforced for the so-called Tier-1 cities like Shanghai, Beijing, and Shenzhen because the economic activity there is such a magnet that unrestricted immigration with the expectation of benefits could swamp those centers.

China calls its system of government socialist, but it has never been a welfare state, mostly because the scale of the population would shred any universal safety net. The last few years have seen an increased public emphasis on family ties. The news is filled with stories of sons who leave for the cities and don't return, while daughters who don't marry and have children before age twenty-seven are labeled "left-over women." The state wants to rescue young people from poverty, then have them rescue their families, because the government simply can't help everyone. If children don't take care of their parents, there isn't much in the way of Plan B, and with the one-child policy having produced millions of single-child families, it only takes one son or daughter shaking off filial piety to leave the older generations stranded. The interaction of the market economy and family expectations *is* the safety net for a majority of citizens.

For the first generation of modern China, opening up to the
world meant becoming the world's workshop, where outsiders
supplied demand, manufacturing instructions, and raw materials. China's only great advantage, an enormous pool of cheap
labor, made the price attractive. Building a market economy that
could lift hundreds of millions of people out of poverty couldn't
happen all at once, though. In the 1970s, the country was too
vast, too poor, and too inexperienced. The first of the trial
businesses set up north of Hong Kong at the beginning of the
opening-up period was employed to dismantle ships for scrap,
since there were not enough trained workers (or machinery,
management, or investment capital) to building anything foreigners would be willing to pay for. From that rough beginning,
China has taken on a larger amount of the world's increasingly
complex manufacturing.

The arc of the Chinese economy has bent from dismantling
ships to assembling simple items for foreign firms to taking on
increasingly sophisticated manufacturing jobs, and all the while
building a domestic market for the growing number of citizens
with at least some disposable income. All of this has happened
at dizzying scale and speed. In the forty years since Deng kicked
off China's opening up and reform, hundreds of millions of people have been lifted out of poverty, and this in turn has made
China not just the world's number one producer, but its number
one consumer for an increasing range of products as well.

The Chinese government is delighted to have most locally
made products find a Chinese market. As the U.S. discovered
in the 1800s, having robust local demand provides a degree of
ballast in an otherwise export-driven economy, and rising local

58   prices and rising local salaries can reinforce one another for a time.

The U.S., with its armies of manicurists and knowledge workers, may have become a post-industrial economy, but we don't live in a post-industrial world. America simply shifted the business of making things five thousand miles to the left. Like every other bit of the world's electronics, mobile phones have long been assembled in China, and like every other bit of electronics, the Chinese have become better at assembling those electronics than any other country in the world. If you want to understand hardware in China, you have to understand how intimate that country's relationship is with the machines that run our lives.

The Chinese Apple

Mobile phones are a funny product, midway between commodity and luxury. They are a commodity in that everyone needs one. They are a luxury in that a phone makes a significant personal statement. Much of this expresses itself as "affordable luxury"—the market for phone cases (which, like everything electronic, reaches its most gonzo instantiation in East Asia) is based on a desire to have a cheap form of individual differentiation of otherwise increasingly identical phones. There have been very few offerings of true luxury electronics other than audio equipment. The only recent version is Vertu, a 2006 spinoff from Nokia, which makes $20,000 phones. (Rubies in the keypad, gold finish, alligator-skin case, and handmade in England, in case you were wondering.) Vertu doesn't even try to argue that the phone is worth $20,000 as a phone—their business model is reminiscent of the old joke about the Russian oligarch who runs into a friend and asks, "Like my new suit? I got it for 1,500!" His friend replies, "You idiot! I know some-place where you could have gotten it for 3,000!" (Last year Vertu announced that it was going to base new phones on Android, further evidence that smartphones are the only phones anyone wants, and that there is no real place for a third operating system other than Apple's iOS and various flavors of Android.)

The phone is now a nearly universal sign of wealth and pref-erences. Someone with a flip phone is either poor or else rich and ostentatiously low tech. Everyone else is selecting between various black slabs of glass that communicate something about both taste and status. In China disposable income and the per-mission to indulge in public individuality are appearing at the same time. Here as everywhere Apple is the brand to own if you are showing off, an imperative the well-off take seriously. Though Apple has a smaller market share than its global aver-age here, there are more iPhone 6s sold in China than anywhere else. Status is a bigger feature of the iPhone here than in the U.S. Electronics stores display phones running Android with the screen facing out, as usual, but iPhones are often displayed case out, to show off the Apple logo. Things that get in the way of that visible status are also an issue; a new kind of phone cases has appeared, after the arrival of the iPhone 6, with a large "6" on the back, so that you don't have to choose between protecting your iPhone and letting the world know you own one.

Apple, here as everywhere, has become the official Best Company. (At a hotel in Zhongguancun, Beijing's tech dis-trict, the lobby gift shop sells Steve Jobs bobblehead dolls.) No Chinese company has done more to try to acquire some of that mojo than Xiaomi. The firm is widely referred to as the "Chinese

Apple," a phrase that carries both a sense of awe at its design prowess and derision at its habits as a copycat (both reactions are warranted). Xiaomi founder and CEO Lei Jun is sometimes jokingly referred to as Leibs, a portmanteau with Jobs. He disputes the comparison, insisting that Xiaomi is closer in spirit to Amazon and Google, but it sticks for a reason. Beyond the fact that Lei wears black T-shirt and jeans at Jobsian product launches, one has to recognize how recently and dramatically the mobile phone market has been transformed by the appearance of the iPhone, and how much that transformation did to move the center of that industry from northern Europe (Nokia, Ericsson) and North America (BlackBerry, Motorola) to Asia.

The iPhone, launched in 2007, singlehandedly altered people's ideas of what a mobile phone could do, transforming it from a thing you used to talk and text with into a powerful, flexible, networked computer that fit in your pocket. In one of the fastest transitions of consumer preference ever, Apple's offering made even the most expensive phones by Nokia and BlackBerry seem like clunkers. I remember running into a designer friend at Nokia as all this was happening. I asked if Nokia's management had an answer to the iPhone. He replied grimly, "They don't even understand the question." Nokia went from being the world's most important mobile phone company to an also-ran in three years, collapsing into Microsoft's waiting arms after another three, a generation of dominance undone in half a decade.

By sidelining Nokia, Apple effectively destroyed what was left of the European mobile phone industry. (Ericsson's phone business had been acquired by Sony in 2001, Alcatel's by the

62   Taiwanese firm TCL in 2005.) This was a huge shift. Nokia
       has eight of the top ten bestselling handsets ever, accounting
       for over 1.3 billion phones. Apple's entire sales output—every
       model of every iPhone ever sold—is smaller than the num-
       ber of Nokia 1100s sold. Apple also dealt a blow to Motorola
       and BlackBerry (the former being acquired by Google, the
       latter in near free-fall). This left the iPhone as a product so
       beautifully designed it occupied a category of one, and with
       competition so hamstrung by historical habits that they
       couldn't even launch an acceptable alternative. What no one, not
       even Apple, understood then is how fast the rest of the industry
       would be transformed.

       Apple had hoped to repeat its success with the iPod and
       have another product category to itself; the nearly instant copy-
       ing of the iPhone has ensured that that will not happen. Even
       as the iPhone was redefining people's sense of what a mobile
       phone could be, Apple had inadvertently created an almost per-
       fect set of conditions for new, heightened competition.

       The shift to smartphones was late in coming to China.
       The combination of cost-conscious consumers and trade bar-
       riers kept the country using ordinary mobile phones for years
       after the iPhone launch. Nevertheless, you could see the coun-
       try was changing, at least if you had the right vantage point. In
       the years running up to the iPhone launch, Lei Jun was drawing
       some hard lessons from his time at Kingsoft. Though the com-
       pany had been in business since the late 1980s, it was always
       a struggle. Like many firms that had tied their fortunes to the
       personal computer, Kingsoft moved from product to product,
       offering everything from a competitor to Microsoft Office to

translation software to anti-virus tools. (There is still a bit of
source code kicking around that Lei wrote in his early Kingsoft
days, designed to kill errant process on Windows computers.)
He tried and failed four times to bring Kingsoft public, suc-
ceeding only on his fifth attempt in 2007. Shortly thereafter, he
stepped down.

While Lei was at Kingsoft, he also founded Joyo.com, one of
China's first online bookstores, which he sold to Amazon.com in
2004 for $75 million. Given today's valuations of Chinese inter-
net firms, that was not a huge payout, but it told him something
about the nature of the market. As he later put it, he learned
that "doing the right thing is more important than doing things
right." When a company is riding a growing trend, management
is under less pressure to do everything perfectly. (Another saying of his from those years is, "Even a pig can fly if it finds itself
in the eye of a storm.") By 2007, the year the iPhone launched,
he understood that PC software of the sort that Kingsoft had
always concentrated on was far from that storm, while anything
that was taking advantage of internet connectivity was right in
the middle. What was not yet clear was how quickly the internet
and mobile phones were going to intersect in China.

Just before the iPhone launched, Google had acquired a
company called Danger that manufactured an innovative but
niche phone called the Sidekick, beloved of internet entrepreneurs and American teens. Danger also had an operating system designed to run phones like the Sidekick, which they called
Android. With the iPhone as visible proof that touch-screen
interfaces were practical, desirable, and affordable, a team at
Google quickly modified Android to use a similar set of tap/

64   swipe/pinch gestures, ending up with an alternative operating system that any other handset manufacturer could use. By 2009, the Android OS was widly successful, and its ubiquity meant that the iPhone would be the niche product among smartphones globally (albeit a fantastically profitable niche). It looked like the essential competition in the mobile phone industry had moved to two companies in Silicon Valley—Apple and Google—separated by ten miles of Highway 85.

This didn't happen, because Google, whose only real hardware expertise was in running servers, turned out to be unable to design and sell phones anyone was eager to buy. (I was an early owner of a Google Nexus and it was … fine.) What Google had done by buying Danger and modifying Android was to bring the old dynamics of the PC market to phones.

In that market, Apple always sold its computer hardware and software together; other than one disastrous experiment, Apple made it impossible for a third party to make a computer running their operating system. Windows computers, by contrast, had never been tied to just one hardware company, which is how IBM, Compaq, Dell, and Hewlett-Packard could all sell Windows machines. After the launch of Android the same dynamics now held true in mobile phones.

The only company you can buy an iPhone from is Apple. Meanwhile hundreds of millions of people were willing to accept Android phones from any number of suppliers as a perfectly adequate substitute, if the price was right. The firms best positioned to pick up these millions of customers were almost all in Asia: Samsung (South Korea), HTC (Taiwan), and Huawei (in Shenzhen, long China's center of manufacturing).

The advantages of East Asian manufacturers are by now familiar: deep expertise in tooling up for novel manufacturing challenges; cheap, skilled labor; investors eager to fund new manufacturing capability; and the nearly pan-Asian leapfrogging of traditional telephone networks by mobile phones. The invention of the first practical mobile phones in Japan in the late 1970s created a business network of firms ready to adopt innovation after innovation in mobile phones, much as the location of Hewlett-Packard in California in the 1930s set the stage for Silicon Valley as a similar network.

These firms have come to dominate the landscape for mobile phones, and for networking gear in general. Huawei became the largest telecommunications manufacturer in the world in 2012, taking over from Ericsson, and continuing the theme of the passage of networking businesses from Europe and North America to Asia. The one place the Huaweis and HTCs faltered in the Chinese market, however, was in personal electronics, which are selected not just on function and price (as, say, routers are) but on style. Once the Chinese had any disposable income to spend on phones, they upgraded from the simplest models to phones as status symbols. Because of the iPhone's price, it was never a serious contender for broad adoption, but the "plain old phone" look of a Huawei or HTC didn't make them huge sellers either. It was Samsung who captured most of the moderately-expensive-and-stylish market as it expanded.

To most outsiders, this market looked saturated. By 2010, China Telecom had agreed to let iPhones onto its network, capturing the price-insensitive part of the market. Nokia still made cheap reliable handsets for the poor, and Samsung took the middle. But the eye of the storm Lei Jun was looking for wasn't centered on hardware. It was centered on the internet.

Xiaomi was one of the first Chinese companies to launch into a country where the internet had become normal. The early successes—Baidu, Tencent, Alibaba—created basic infrastructure where there had been none. By 2010, though, the internet had become a background assumption in almost everyone's calculations. Every age has its ideal invention—at various points, society has decided that Roman roads or pendulum clocks or the steam engine exhibit characteristics that exemplify human achievement in ways that become metaphors for society. A business is said to run like clockwork, smuggling slaves to freedom is likened to an railroad, and so on. For the last thirty years, software was made to use metaphors from the real world: Your computer was said to have folders; tablets use page-turning motions to move through files; search is indicated by a magnifying glass (though at a scale where a telescope would be a better metaphor). These old patterns were trying to get people who'd only ever grown up with physical interfaces to feel an illusory sense of comfort about our digital products, whose enormous complexity and configurability were squashed down into a set of essentially nostalgic icons, a pattern called skeumorphic, referring to design elements that mimic earlier forms.

Recently, however, the ideal invention has been the network. The rise of networks—ranging from the cellular to the social—has been accompanied by new tools for understanding how networks actually work. After we'd had about a decade to get used to the web, we began to import expectations from a software-filled universe back to the real world. We're now

seeing the reversal of skeumorphic patterns. The first generation that grew up with digital affordances as a birthright now
expects its physical objects to exhibit the same properties.
Toddlers who grow up playing with a parent's phone are often
puzzled when they can't swipe the image on a TV. Print layouts
emulate web pages. The scholar Ryan Calo notes that the Google
campus has adopted the legal logic of "clickwrap," the disclaimer
that you falsely aver to have read when you download new software or content. The company's Mountain View headquarters
announces on a screen that by walking onto the campus, you
consent to its non-disclosure agreement.

What's coming is the re-engineering of every complex
object to behave like software. This pattern is sometimes called
the Internet of Things, though it's really the regular old internet, just one in which most of the world's objects have computers inside them. We're almost there today—every device
you own with a battery also has a chip in it, with the possible
exception of your flashlight. (Test your batteries, by the way.)
In a world where any object more complex than a sofa has at
least some computing power, you can change the way your
home looks or works by acquiring new software, rather than
by buying appliances. This is the logic of most computers, of
course, but it is most of all the logic of the smartphone, where
downloading a new app can turn it into a very different kind
of device.

For Xiaomi, that transformation is a weekly event. Every
week, Orange Friday brings slight tweaks to obsess over, including, critically, the improved performance and battery life that
were key to making early versions of MIUI popular.

68   In simple environments, people acquire new capabilities by acquiring new objects. In complex environments, people acquire new capabilities by using new services. And as anyone who's busily added apps to their smartphone knows, the one problem apps are no good at solving is the problem of having too many apps. Environments with rising complexity always create business opportunities for third parties to step in and manage that complexity. One form of success in a complex environment is someone who consistently pays you to keep at bay the very complexity you cause. Achieving that state is Xiaomi's long-term goal. As Xiaomi investor Richard Liu puts it, "We never care about number of phones sold. We care about number of users converted." The company had effectively no profit in 2011, and an astonishingly low profit margin of 1.8 percent in 2013, plowing almost every yuan back into growth. (Having raised $1 billion at the end of 2014, the company is now set to start generating real profit.) This idea—head count matters more than revenues early on—is an old internet pattern, where business after business first scaled up and only then pursued revenues, a pattern followed by Yahoo, Google, and Amazon.

Xiaomi has always been focused on maintaining that pattern even when it moved into hardware. Selling a Xiaomi phone generates some income, but more importantly, it becomes a way to distribute the MIUI interface. The user experiences MIUI as a way to operate her phone, but it is actually a bundle of potential new services. User engagement is Xiaomi's founding logic. When they began, with a small group of employees and just enough cash to get them through early milestones, they began recruiting. This is the normal case, but for Xiaomi recruiting wasn't just about new employees, they also began recruiting new users. This is one of many unusual things the company did—recruiting users of a product that doesn't exist yet seems backwards. Yet those users, or at least potential users, were critical to the speed with which the company would be able to build MIUI, its core product.

Given that the firm was founded by some of the best technical minds in China, it's easy to wonder what Xiaomi got out of these first users. What could they have access to that the people inside the firm didn't know, given that they were building the software? The simplest answer is that the user had access to reality—every company builds a bubble around itself, where the products get built and tested in a more controlled environment than they get used in. This is especially true of complex software. What the early users enabled Xiaomi to see was how MIUI actually worked when real (albeit unusually technically proficient) people tried to install it on a wide variety of devices.

This pattern continues to this day. Xiaomi's marketing chief Tony Wei says that new software sent to early testers will generate thousands of reports back overnight, and this openness in turn allows them to try out, test, and fix the critical source of their commercial advantage, a version of the operating system that lets them tie their own services to the user's device. These are not mere bug reports, of the sort most software now generates automatically. These are user reviews, questions not just about the technical aspects of MIUI but about which features the user likes, dislikes, or wants to see in the future. (The fever users generate more technical feedback, the flood users more emotional feedback.) For a company dedicated to creating cheap products, it says quite a lot about their strategy that they run their own call centers; they consider that experience too important to outsource. The users generating these reports were essential to the company's start, but unlike many internet firms, which build a product with early users and then displace them as the firm grows, Xiaomi has never forgotten the earliest users, in private or public. In March of 2015, when the company had 100 million users of MIUI, their understandably triumphal press release emphasized the work of the original hundred in the first paragraph.

They have managed to extend this sense of importance to their other users. As the number of MIUI users began to include non-geeks, Xiaomi began to behave in ways that would generate interest and loyalty in the general public. Their most famous technique is their flash sales, where a limited number of devices are sold at an announced time, and potential customers have to line up (online) for the right to try to purchase one of the devices. Flash sales generate the "sold out in minutes" figures so often associated with the company. In a recent move to sell phones in India, they sold 40,000 of their cheap RedMi 1S phones in *four seconds*. Here again, as with many of their later moves, the company understood that this was where they needed to be early on. Lei Jun had a great appreciation for the odd dynamics of ecommerce from his days at Joyo, where the peak number of transactions a site can be asked to perform is a significant multiple of the median load. When testing an early version of their ecommerce platform, the company opted to sell cans of Coke to the employees. The original plan was to sell the Coke for 1 yuan (about 15 cents), but Lei decided on a more radical strategy—sell a Coke for 1 *mao* (1.5 cents). Demand for 1 *mao* Cokes subjected the system to enormous load even with very few users, exactly the capabilities they'd need to build up to handle flash sales.

In one of the company's Beijing offices, there is a case displaying fetish object versions of the company's products. Users will draw the Xiaomi logo and draw or trace the outlines of the interface on cardboard or wood, and then send these to the company as tokens of appreciation. The most remarkable version I saw was a block made of thousands of individual millet grains glued together into a phone-sized brick. (Millet looks, well, it looks like little rice.) This particular Mi Fan had then painted company's logo and basic UI on the front in black, a task that had taken three nights.

Another pattern Xiaomi adopted was the stack, which is how programmers describe the notional set of layers that make it possible to keep different kinds of software separate enough to manage them. On the internet, the lowest layer of the stack, by convention, is the hardware that makes up the networks over which messages run. It could be Wi-Fi or a plug in the back of your computer, and the medium of transmission could be copper or glass or air, but all of those details are hidden ("encapsulated") from the users. If you've ever used the internet in a moving car or train, you can appreciate the value of treating network access as a set of interchangeable services—your browser stays connected, even as you pass in and out of multiple wireless networks.

As you rise up the layers (the canonical model has seven, like the Hindu layers of heaven, but practical instantiations have fewer) you get closer to where the people are, with messaging

72   layers and presentation layers. All computing and network environments run this way—an iPhone separates hardware from network interface from operating system from apps just as surely as a Samsung or Xiaomi does. What makes the latter's strategy special is how they use the layering.

Prior to Xiaomi, phone manufacturers treated the operating system as the thing they needed to make the hardware valuable enough to be worth paying for. This recapitulates the early state of personal computers from the late 1970s, where the value was assumed to be in the physical computer, and the operating system was thrown in. In the same way Microsoft realized that as PCs became a commodity, the money would be in software, Xiaomi realized that since phone hardware could accept multiple different operating systems, it could get its operating system onto smartphones without making or selling any hardware at all. Once it had a partially working product, it could ship it as a ROM—"read-only memory," a set of basic and unchanging instructions that live on the phone. Its hundred initial users and their immediate followers would then take these ROMs and install them on their Motorola and Samsung and HTC phones, using a process called side-loading best described as "don't ask." Though layered interfaces between hardware and software have been the normal case in the industry for half a century, and though the largest fortune in computing history was made by selling the operating system unbundled from hardware, Xiaomi was the first company to really invest in that trick with mobile phones.

But as the MIUI improved and spread to more users (each effect driving the other) it became apparent that downloading and installing a new operating system would always be a minority pursuit. When you start with a hundred users, getting to a thousand is a 1,000 percent increase, torrid growth on a tiny base. A common rule of thumb among the makers of consumer electronics is that something like 85 percent of users will never change any of the factory defaults. (This was the source of the blinking "12:00" on millions of video recorders.) Given this, the maximum conceivable size for the Xiaomi market was 15 percent of the total, and even then they had to convince those users to actually download and install MIUI. At some point, long before the millions of users they hoped for, the "install our operating system on your phone!" strategy was going to stop working.

Like the Tokyo Tsushin Kogyo company of post-war Japan, which changed its name to Sony to reach a global audience, Xiaomi has always aspired to global identity (as with the very expensive acquisition of Mi.com). The company's first outfit outside of China was in Singapore, which has a technocratic government like China's, but one with political parties and much greater tolerance for free speech. (It is also a fraction of China's size, a city-state with 5.3 million people.) Singapore was more of a base of operations than a big second market. From there, the company has expanded to Malaysia, Indonesia, and, the other big prize, India. Their Indian launch is an indication of the sorts of problems Xiaomi, and Chinese firms in general, will face when expanding. Their initial opening in India was marred by a lawsuit from Ericsson, which forced them to stop selling phones while negotiating access to Ericsson's patent pool. After sales were suspended, an online outfit named XiaomiShop.com kept selling the banned phones in the country. When Ericsson complained, Xiaomi pointed out that it had nothing to do with XiaomiShop.com, which was buying phones online in China and re-selling them over the Himalayan border.

India is an economic basket case compared to China. GDP per capita is between two and four times higher for China than India, depending on how you measure it, but India's economy is growing faster than China's, and there's no guarantee that Xiaomi can pull off the "cheap phone for design-conscious users" trick in India. Narendra Modi, India's prime minister, has publicly called on Xiaomi to manufacture phones in India, while Indian firms like Micromax and Lava are trying to copy Xiaomi's methods for their local market. (Micromax just unseated Samsung as India's number one phone vendor, another page from Xiaomi's book.) The cell phone market in other countries matters, of course, but not like India and China matter, the only billion-strong markets in the world. Xiaomi's expansion is both predicated on, and a test of, what Chinese manufacturing will look like as it incorporates high-quality industrial design, not just industrial production.

Industrial design is just part of Xiaomi's goal. The company has consistently set its sights on being a services company, and online services seem like an obvious candidate for globalization. The internet is the first communications medium where international transmission generates no additional charges to users; an email crosses an ocean as easily as it crosses town. This ease of adoption is one of the principal advantages U.S. internet companies have as they expand. The Facebook of India is Facebook. The Facebook of Australia is also Facebook. The Facebook of China, however, is Renren, launched

in 2005. (The Google of China is Baidu, and the Twitter of China
is Sina Weibo.)

These Chinese sites have all become behemoths, many
of them having had initial public offerings earlier this decade.
Their original competitive edge was language—at a time when
U.S. internet firms were not as aggressive about international
expansion, and when translating sites into multiple languages
("localization") was more expensive and time consuming, a
native Chinese-language site had a cultural advantage.

A second, technical advantage has appeared over the last
half-dozen years, as the Chinese government realized that social
media creates a space for political conversation—and, even
more alarmingly, for coordination of real-world action. After
observing the use of social media during the Iranian uprising
of June 2009, and during riots in Ürümqi, in their own Xinjiang
Uyghur Autonomous Region the following month, they
began blocking such sites from the West, especially Facebook
and Twitter.

They also decided that any user-generated content—words
and images and, increasingly, video—should be hosted in the
country, by Chinese firms, to minimize outside influence and
maximize Beijing's oversight. (Last year, a potential Facebook
investment in Xiaomi was sidelined in part because of the
political ramifications.) This combination of blockade and local
control is what allows them to prevent the Chinese population
from seeing images like those sent from thousands of camera-
phones during the huge pro-democracy rallies in Hong Kong in
late 2014. The government simply blocks foreign photo-sharing
services, and they direct local apps to delete any such images

76   that appear. They never succeed completely, but since their goal is to prevent the majority of the Chinese people from seeing the images, they don't need to keep them away from every last person.

That blockade has remained in place ever since, and has been expanded to include many other sites useful for publishing or sharing content: YouTube, WordPress, Instagram, WikiLeaks, and so on. As the rest of the world has converged on a small set of large-scale social media sites, China is the only country that has a robust but alternate set of social media on offer to its population.

The Chinese internet service companies that have flourished in the last decade are often run by people who have studied in the U.S. and have global ambitions; given the relative ease of offering an internet service anywhere, these companies seem likely targets for export. However, unlike manufacturing firms, internet service businesses in China face serious challenges to globalization. The first challenge is, again, language. The vast majority of people who can read Chinese live in China, so any Chinese site has already reached most of its potential audience. By contrast, speakers of English as an additional language outnumber native speakers five to one, so any site that launches in English, in any country, still has more potential global users than local ones, even without translation.

The next challenge is commercial. The Chinese banking system is notoriously disconnected from the rest of the world (as anyone trying to use a credit card outside nice hotels quickly discovers). Any site optimized for handling online transactions in China has invested in an expertise almost as non-portable as the language. Similarly, the relatively lax enforcement of intellectual property claims in China means that expanding out of China requires not just leaving a familiar market behind, but abandoning a key set of governmental protections as well. Alibaba, owner of Taobao and Tmall, China's two largest ecommerce sites, sold off 11 Main, its attempt at a U.S. site, after less than a year; being the dominant Chinese firm did not prove a transferrable advantage.

Then there is the challenge of perception. As reports of Chinese hacking of foreign governments and businesses continue, and as the Chinese government is moving from foreign suppliers of networking equipment, like Cisco, to local ones, like Huawei, there is increasing worry on the part of foreign governments and citizens about using any Chinese-hosted service.

And finally, there is the challenge of operating under an authoritarian government. The blockade creates an environment where local firms can flourish; but like all protectorates, it can also subject the protected to abuses of power. In April of 2015, China launched a cyber attack that was quickly dubbed "the Great Cannon." The attack targeted GreatFire, a U.S.-based service that both tracks and fights Chinese censorship. The Great Cannon hijacked incoming traffic and re-directed it across the internet, to flood its target with useless requests. Much of the traffic it hijacked came from users of Baidu, China's search engine giant. Even though Baidu is a favored brand within China and enjoys a good working relationship with the government, when it came time to mount an attack, the government was willing to tarnish Baidu's brand on the global stage to further its political goals.

As Xiaomi tries to export its service businesses and not just its hardware, it has faced some of these problems already, and

78 will face more in the future. The company has invested heavily in its English communications. Hugo Barra, the executive who came over from Google, is the company's face in much of the world, and Lei Jun recently appeared speaking English at the Indian launch of the Mi4i phone. At one point, trying to whip up the crowd, he began shouting, "Are you OK? Are you OK?" sounding more like a paramedic than a CEO. He was affectionately mocked for his English, with videos showing up on Youku (Chinese YouTube), amid general approval that he was trying.

They have also largely overcome the commercial optimization problem, in part by investing heavily in multi-country versions of their ecommerce site, Mi.com, and in part by choosing to work with local sites in those countries, like Flipkart in India, in order to reach more citizens.

The perception problem is harder, of course, as they learned with their missteps in sending data from Indian phones to their servers in Beijing. This problem is likely to get worse, as the administration of Xi Jinping continues to crack down not just on social media but on anything that supports civil society without direct party oversight, and as the cold trade war between China and the U.S. over electronic hardware and software deepens. And the threat that the government will do something to damage the perception of Chinese internet service businesses (or of Xiaomi in particular, as they did to Baidu) is constant.

By linking hardware and software, Xiaomi is trying to bridge the gap between easy-to-export objects and hard-to-export services. Lei Jun has ambitions for his firm to be like Sony under Akio Morita, who wanted to change the connotations of "Made in Japan" from cheaply manufactured products to quality ones,

but this will mean emphasizing the Chinese reputation for production while avoiding its reputation for authoritarian control.

Because the founders designed the firm with globalization in mind from the beginning, Xiaomi has the best chance
to do this of any Chinese services firm. "Xiaomi's mission is to
change the world's view of Chinese products," Lei told the *Wall
Street Journal* during the Indian launch of the Mi4i, which sold
40,000 units in fifteen seconds. (It helped that the phone is a
mid-range, scaled down version of the Mi4, and sold in India
for 13,000 rupees, or about $200, compared to the iPhone 6 at
$800.) As the newspaper noted, India is an important step in
realizing Lei's global ambitions to make the first Chinese brand
that's "cool" abroad. The balancing act for the firm is whether
they can balance the dictates of being a cool global firm and a
loyal Chinese one.

Maker Movement

China is the world headquarters of making things. Golan Levin, who created a twenty-first century upgrade of the camera lucida drawing tool using Kickstarter, decided to have his product made here. When I asked him what he'd learned bridging the gap between American maker culture and Chinese manufacturing, he replied, "The hardest thing to understand when talking to Chinese manufacturers is that there is no shelf. They'd ask, 'What sort of screw do you want here?' And we'd say, 'Well, let's see, what do you have off the shelf?' And they'd ask again, 'Well, what do you want to use?'"

Levin said it took them a couple of go-rounds before they realized that there wasn't any shelf to get things off of, that any given screw was going to be as cheap as any other because none of those screws existed in advance of demand. The producers didn't own screws, they owned machines for making screws,

so you might as well design everything from scratch. He had
gone so far up the supply chain there were no more supplies.

We rely on our electronic devices to augment almost every
activity we undertake, and yet we've all become alienated from
the way that technology gets made. Despite the promise of Intel
Inside™, few of us have ever seen a chip, motherboard, graph-
ics processing unit, or hard drive. Computers come from stores;
none of us build our own computers anymore, and none of us
know anyone who does. Mike Daisey's long monologue about
Apple and China, *The Agony and the Ecstasy of Steve Jobs*, was iffy
journalism but brilliant storytelling, tracing his realization that
people have made every bit of electronics he owns. In 2011, an
iPhone was sold in the U.S. with test photos still on it, taken by
a worker inside Foxconn, the giant electronics manufacturer in
southern China, one of the rare public traces of the human effort
that goes into making.

In any big Chinese city and most of the medium-sized ones
(which is to say any of the hundreds of cities here with a population larger than Seattle's) there will be a big electronics mall—
Cybermart is a common brand, though there are many others.
These are multi-story stores, divided up into booths of different sizes, like a trade show, each rented out to a different mer-
chant. To an American eye, the whole thing is a little nuts, but it
works, and traveling through one is like taking a core sample of
electronics in China, only here you start at the bottom and work
your way up.

The economics of retail floor space mean that the showy,
high-margin stuff is at street level, and the gritty low-margin
goods are at the top. The ground floor of a Cybermart will tend

toward the bright, clean, white space of a showroom floor. There will be a few big booths selling premium brands like Samsung, Apple, Sony, and Lenovo. You could be in a Best Buy in Ohio, except for the Chinese signs. This is the least interesting part of the space; the real action is upstairs.

As you go up a floor, you'll see more merchants in smaller booths. (Chinese electronics retail remains a distinctly small-scale affair, often literally mom and pop, with the kids hanging out in the store after school.) Up here, there will be more off-brands, Acer and Asus for laptops, Oppo and Huawei for phones, as well as flip phones and Nokias on their way to obsolescence. There may be some nominally-banned-but-who-are-we-kidding consoles and games like the Nintendo Wii. There will be lots of accouterments—mice, joysticks, chargers, and cases. There's usually a soup place up here, for the workers, and maybe a FamilyMart or a 7-Eleven. Every now and again, some new product will come along, and half the booths will try to sell it. Last fall it was laser pointers. This spring it's USB chargers with embedded blinking LEDs. Unlike the clean well-lit booths that tend to cluster at street level, the upstairs vibe is RadioShack with Chinese characteristics.

At the back of this floor (or, in bigger ones, on the next floor up) you'll change environments again, with booths selling something most Americans haven't seen in two decades: parts. Want to make your own PC? All the fixings are here—chassis, power supply, motherboard, RAM in any pin configuration you'd care to name, enough hard drives to store the world's vacation photos. (When was the last time you laid eyes on a heat sink? Maybe never? They're available by the crateful, and so cheap.)

Nothing here is branded Dell or Sony; many things are branded
Nanya or Western Digital, the companies that make the pieces
that make your Dells and Sonys worth owning. China is a place
where people still build their own computers, a habit Americans
dropped about the time Windows 3.1 shipped. The vibe in this
part of the store is more warehouse than retail. There's little
advertising or style—the customers up here know what they
want and are shopping on price.

Alongside the unfamiliar sight of computers *in potentia*,
there will be an unfamiliar smell, smoky and metallic. That
is the smell of solder, the molten mixture of tin and lead that
wires all of your electronics together. The people with the sol-
dering guns are engaged in full-on disassembly, modification,
and repair of pretty much anything that has a screen. (Anyone
in a city with a large Chinatown can probably find some small-
scale version of this.) You'll see broken screens everywhere in
this part of the store, enough shattered glass to redo the façade
of every Urban Outfitters in America. You know that message
on the underbelly of your electronics? "No user-serviceable
parts inside"? They scoff at that message here. This part of the
Cybermart, far more than the sales floor or even the booths selling raw parts, exemplifies the Chinese relationship to electronics as a completely ordinary thing you make and modify, free of
any trace of mystery.

This is in strong contrast to the new vogue for making
things in the U.S., often called the Maker Movement. Maker-
y-ness in the U.S. comes as part of a complex of oppositional
attitudes toward mainstream culture that is more about social
signaling than unvarnished commitment to DIY. The Maker

84   Movement involves ostentatiously DIY products, designed and assembled against a background of nostalgia for the old U.S. manufacturing industry, often produced in small batches for connoisseurs of the handmade, created as a form of conspicuous production.

Meanwhile, back in China, nothing in the Maker Movement is taking place against a background of nostalgia, because "the time when this country knew how to make things" is just a synonym for "this morning." Homemade stuff here is generally homemade because one, it's cheaper or because two, the thing you need doesn't exist. The part of Chinese manufacturing culture that says, "Hey, here's a pleather purse branded YSL" is prevalent but boring. But away from the Fauxlex watches and Svmsmvg phones, brand still means what it did in the U.S. a century ago, a sign of quality in an environment with a lot of lousy versions of everything floating around. A company can do well by making moderately more expensive but considerably more reliable vacuum cleaners, as long as it can communicate that fact to people who need vacuum cleaners. Overt signs of quality have been hard to come by until recently—high-end milk advertises its arrival via a 4-degree-cold chain right there on the carton and the best Chinese appliance firm, originally the Qingdao Refrigerator Company, renamed itself Haier, so that people would associate it with Germany.

Having a brand that stands for quality at an acceptable price is a big deal here, because this country is cheap, cheap the way many Americans were if they grew up in the Great Depression. Last fall, I went by my local Cybermart to get a Mi4, then the hottest Chinese phone since the Mi3. I'd gone there straight from a meeting so I even looked like a businessman instead of a nerd, and I rolled up to a second-floor booth selling Xiaomis and got the attention of the lady behind the counter. I am a bald white guy, I speak pidgin Mandarin with a flat Midwestern accent, and I was in a suit—the only way I'd have looked like an easier mark is if 100 yuan notes were spilling out of my pockets like in cartoons. I announced that I wanted to buy the priciest phone any Chinese company has ever produced. The lady behind the counter looked at me and said, in English, "You don't buy that phone. Too expensive." China is cheap like that: Even people paid to take your money are offended if they think something costs too much. Tell me the next time that happens to you at Best Buy.

Thin wallets are the mother of invention; so much design in China is about saving money. The cheaper a phone is, the likelier it is to be dual-SIM, so you can save money network-hopping. Electronics markets are filled with every type of video and audio and network switcher and splitter and converter because people don't throw the old stuff out, they just find a new place for it to work. This profusion of systems—high tech and low, custom-built and jury-rigged, all side by side, is the normal case here.

This widespread competence in making is part of what made Xiaomi possible. The move into manufacturing the Mi1 came in 2011, in part because of frustration that Xiaomi could only get so much value out of optimizing their software for other people's hardware, and partly because the geek love that sustained them in their first year was not evenly spread through the world's population. (When was the last time you replaced the operating system in your phone?) Manufacturing was a huge shift for a firm that had been, up until that point, a software

86   company, including a new round of hiring, since most of the original co-founders had been software and services people. Indeed, Xiaomi was seen as such an unlikely target for releasing hardware that the Chinese rumor mill assumed Xiaomi would partner with an existing hardware firm that wanted to enter the Chinese market, like Motorola or LG, in order to offer a phone pre-loaded with MIUI.

The arrival of the phone was a surprise, and it was greeted with skepticism. As one analyst said at the time, "The market for the device is very narrow since it will only cater to customers in first-tier cities. I think the company is too naive about the cell phone market." Even people paid to follow the Chinese mobile phone market did not understand how quickly or completely consumer preference had changed. Hundreds of millions of people were in the market for a good enough, cheap enough smartphone.

Unlike software, launching a new hardware brand presents a chicken-and-egg problem—manufacturers don't want to sell to you if you don't have customers, in part because every bit of contracted production creates both a risk of not getting paid and a risk that you will have promised a key bit of inventory (screens of a certain size; memory chips of a certain volume) to several small manufacturers when Samsung comes calling. Then there are the costs of dealing with several small companies at once, versus a few large ones. These difficulties are in turn reflected in the cost of hardware. A firm buying a hundred screens, to create test models, will pay something like $35 per screen, but a firm buying a hundred thousand screens will only pay $20. This creates a barrier to entry for new brands; the wholesale vendors of electronic parts have a strong preference for having a few large contracts from a few large firms, versus many small contracts from many small firms.

Xiaomi's response to this problem was to lean on their expertise in online commerce, coupled with the potential size of their initial market. The problem small, new firms face is uncertainty—forecasting demand is a famously hard problem for retail. Because of its experience with Mi Fans, Xiaomi was able to pre-sell phones it had not made yet, pocketing the deposit money to reassure its suppliers, and negotiating for parts with visible demand behind them. Online selling allowed it to buy components as it went along. (The constant upgrading of electronic production means that some components fall in price by 1 to 2 percent a *week*.) In addition, working online allowed phones to be made in batches, rather than having to have them made continuously. This allowed Xiaomi to make and sell only the phones it could get components for.

All of this allowed Xiaomi to use its original market as a staging area for early growth before becoming a global firm. The population of mobile phone users in China is larger than the combined population of the U.S. and Western Europe. (Not larger than the mobile phone using population, larger than the *total* population.) Nearly 90 percent of the adult population in China has a mobile phone (the only kind, in many households). The rise of Xiaomi marks another phase of maturity in the Chinese market, from simply trying to escape penury in the 1970s and 1980s; to seeing local markets for cheap goods grow by leaps and bounds, even as the stuff for export was better made and more valuable; to today, when a Chinese firm can

88   make money selling products designed to do well anywhere. Lei
Jun seems determined to demonstrate that China is more than
cheap factories and knockoffs—that a Chinese brand can even
be coveted and adored. That this transition has happened at all,
given the parlous state of the country only a generation ago, is
little short of a miracle.

The Chinese Dream

It is difficult to describe how poor China was at the end of World War II, the period when Western nations began the three decades of growth that put our economies ahead of most of the world. Shortly after the war ended, Mao Zedong's army finally chased the nationalist government out of mainland China to shelter in Taiwan. Eager to catch up and a great believer in the planned alternative to market economies (it seemed like a good idea at the time), the Chinese Communist Party turned its attention to rebuilding a country that had suffered two decades of civil war, interrupted by a horrific Japanese invasion.

Mao was a better military commander than a peacetime leader. Having taken over a desperately poor country and eager to make progress, he made two fateful decisions at the outset of his rule. The first, widely known, was the Great Leap Forward. This was a set of national policies implemented in the 1950s that included collectivization of agriculture, a disaster everywhere it has been tried, but nowhere as much as China. The resulting famine killed between 20 and 40 million people in three years, the deadliest in human history.

A second, less well-known decision was based on his simple calculation "More people, more power." Copying a Soviet system of the same name, Mao created policy preferences for Hero Mothers, women who had many children. At a time when much of the rest of the world, including most of the developing world, saw reductions in population growth, China's average remained at around six children per woman. Over the next two decades, China added the population of South America, even as they'd hampered their agricultural system.

Mao's last great program was the Cultural Revolution of the late 1960s, where ideological purity rather than competence became the rationale for promotion at state-owned (which is to say substantially all) Chinese industry. Colleges and universities were closed, while teenagers were released from school and dispatched to attack the Four Olds: Old Customs, Old Culture, Old Habits, and Old Ideas. (While touring a cave in Hangzhou filled with Buddhist carvings, I noticed that the figures near ground level had all had their faces smashed off, while the figures higher up the walls were intact. When I asked our guide about this, she said simply, "The high school students did not bring a ladder that day.")

And then Mao died. It wasn't unexpected—Parkinson's had been eating away at him for years—but he was so manipulative, right up to the end, that there was no orderly transition in place. The country was left poor, crowded, and without competent workers or management, and an educational system that had ceased to educate. A genial second-string politician took over for a couple of years, but it was really Deng Xiaoping, already old and going deaf, who took over. Deng's job was to dismantle the worst aspects of Mao's China, and there was in all the world only one wrench large enough to fit around that problem: the market.

Many of the things that homogenize rich countries—phones, broadcast media, electricity—have only been constructed on a national scale here in the last generation. For the Chinese, their regional highway system is something that appeared recently, while the current building boom is without parallel or precedent. China used more cement in the first three years of this decade than the U.S. used in the twentieth century. (Much more: 6.6 billion tons to 4.5.)

But the advent of the market has also brought about immense economic disparities. Tianjin, a metropolitan province just east of Beijing, has a GDP per capita four times that of the mostly rural province of Guizhou, a gap more than twice as large as that between Connecticut and Mississippi. I work at the eastern edge of Lujiazui, home of any Shanghainese skyscraper you've ever seen photographed and ground zero of flashy malls, including the unsubtly named Super Brand Mall. Shanghai is The Future, the backdrop of science fiction films like *Her* and *Looper*, with landscape-scale buildings that double as televisions, subways charged to double cracks, and the world's only commercial maglev train running out to the airport every fifteen minutes. (Innovation in transportation is a consistent theme; one of my neighbors commutes to work on a motorized unicycle.) At the other extreme, when my friend Kevin Kelly travels inland, he says he orients himself to the degree of technological

92   sophistication for any given place by looking at how much metal is around. There are still places where there may be a few metal tools, but the buildings are constructed entirely from stone or wood.

Against this inequality, however, there is a Chinese dream, shared by people all over the country. Chinese governments have always trafficked in the rhetoric of national greatness, even when weak. (The name of the country is often translated as "Middle Kingdom", but "Central Kingdom" is closer to its sense in English, the country at the center of the world.) What makes the contemporary Chinese dream unusual is that it is also an individual one, shared by hundreds of millions of ordinary citizens. It started in the 1970s as a dream that maybe the hard times were over, that the catastrophe of the Great Leap Forward and the madness of the Cultural Revolution could be set aside, and that ordinary people could not only get on with their lives, but that those lives might become easier. This personal part of the Chinese dream is very much like the American one, down to the obsession with housing, and tied to the aspirations of anyone in a market economy: If you work hard, your life will improve, and that improvement will include material comfort and ownership of a home and a car.

The arrival of the market in China has been a humanitarian triumph. Poverty, endemic in Mao's China, has plummeted in a single generation, from 84 percent in 1981 to 13 percent by 2008. For hundreds of millions, the Chinese economy has progressed from providing bare subsistence to broad comfort. As the middle-class ranks swell, there is a market for anything anyone has ever needed or wanted in an industrialized country, from shoes to air conditioners, and it is a very large market; if you make something that appeals to 5 percent of the Chinese population, you have a potential market the size of France.

Because transportation networks were so bad by the mid-1970s, individual towns had become adept at making replacement parts for the machines they used, because trying to buy parts and have them shipped was so often hopeless. These "town and village" enterprises, as they were called (and the source of the *shanzhai* label for locally made, somewhat thrown-together technology), became the first success story of modern China, the first glimmer of the export-driven powerhouse we know today. As the post-Mao government worked to get the railway network moving again (at one point, China's supreme leader had on his to-do list synchronizing the lunch breaks of rail workers), town and village enterprises began to be able to offer their wares outside their immediate region.

Production plus transportation makes for two-thirds of an economy, but there still needed to be customers, and there were few to be had in China. There was no domestic market to speak of. The state both requisitioned and produced the materials needed for heavy industry, while consumer demand was for the bare necessities like food, clothes, and housewares, all of it cheap in every sense of the word. Deng is said to have been astonished to discover, decades after the communist revolution, that the average Chinese family couldn't afford a radio, and residential electrification of Chinese towns—the *towns*—was only completed in the 1990s.

If you want a two-slide comparison of the difference between the local U.S. and Chinese economies, you could do

94   worse than comparing in-flight catalogs, commercial day-dreams on glossy paper. On U.S. airlines, the ridiculous SkyMall, recently bankrupted by the FAA's decision to let people keep playing games on their phones as they land, was a glimpse into a market where everyone who could board an airplane already had everything they needed. So what did SkyMall invite Americans to daydream about? An outdoor chaise longue, for dogs. A mirrored door-mounted jewelry armoire. A spatula with an LED flashlight in the handle. SkyMall was shopping for a culture whose middle class thinks, "Hmm, I could use a Tetris lamp, and maybe some rechargeable heated slippers." China's daydreams are different. Here's what's on sale in Chinese in-flight shopping magazines: rice cooker, baby thermometer, skillet, vacuum cleaner, iron, paring knife, umbrella. It's like Sears and Roebuck with QR codes. This is shopping for a culture whose middle class thinks, "Hmm, I think I need a set of spoons, and maybe a toaster oven."

The ordinary trappings of the middle class have only become widely distributed in this century, a process that it still going on. My family moved to the newer part of Shanghai, east of the river, and many of the young parents we see out in our neighborhood were the first generation in their families to go to college, and also the first generation to own a non-stick pan, or buy their kid a bike with training wheels. In the U.S., many of the changes we associate with comfortable middle-class life occurred over the course of several decades after the end of World War II. This country simply has a different way of being a modernizing society, and any American who comes here to do business becomes an intuitive Sinologist, if only to understand the basics of getting

along. In the big Chinese cities, rising income, urbanization, pro-
fessionalization, education, creature comforts, mobile phones, and social media are all arriving at the same time. There are days that feel like 1955 and 1995 and 2015 during the same afternoon.

These facts—the starting position of abject need and poor infrastructure, coupled with the country's scale—combine to give the Chinese economy a curiously spring-loaded quality. Trends here often start later than in the rest of the world, but when they move, they move faster. You can see this in Xiaomi's bet on smartphones: At a time when smartphone penetration was under 10 percent, Lei Jun and his co-founders could see the way the country was going. By 2012, just a year after they launched their first phone, Lei said, "I expect two-thirds of Chinese people will be using smartphones by 2013," as rapid a shift in adoption as has happened in any country. (He was right.)

This is the Chinese dream, a faith on the part of middle-class citizens that their efforts will be rewarded, and faith that they will be able to keep most of those rewards for themselves. China delivers the thing that most citizens want in most countries most years: a sense that life is getting better. The second part of the dream, the part that is in the background of the American dream but the foreground in many places here, is the link between personal success and national greatness. This is the Chinese Dream, upper-case, and it is now being pushed by the current government on billboards and in public pronouncements all over the country.

Beijing wants a country whose citizens enjoy a high degree of economic freedom, a high degree of personal freedom, and a low degree of political freedom. This is a strange mix, and a novel one. While communism's pedigree stretches back to the early 1800s, a government that offers its people a dynamic market and great personal choice over how they live their lives, in return for political quietude, is an invention of the late twentieth century. Previously, the purchase of complacency has only happened in countries with extractive wealth, and particularly oil. Autocratic countries without extractive wealth have almost all used visible public threats as a way of keeping the citizens out of politics. (This was the communism of Romania and Yugoslavia.) China, by contrast and almost uniquely, uses a vibrant market to convince the citizenry that their views of governance are both unneeded and irrelevant.

This market-supported bargain has worked better than almost anyone expected, but the days when the rising tide really did lift all boats, and where the economic tide was rising consistently quickly, are now ending. The Chinese Dream is Xi's attempt to deal with the end of the easy growth. The moderately prosperous society he is proposing to (comprehensively) build is a way of trying to deflate the rising expectations of the middle class for both marked economic and political improvement. These are all entries in his longer-term goal of bringing China's single-party system into some sort of self-governing norm, while at the same time convincing the Chinese masses to accept a slowing economy and significant income inequality.

The Chinese Dream, as an elaborate form of national pride, appears just as the older Chinese dream of widespread economic improvement is ending. As a Chinese businessman born in the 1970s said to me last year, "We are lucky. When we were young, China needed everything, so anything we sold, we made money. For young people today, only the internet is still like that." The moderately prosperous society is the consolation prize for living in a China that will mint many more millionaires and billionaires, while economic improvement for the masses is slowing down. It was easy enough for Deng Xiaoping to say that it was OK for some people to get rich first, when even senior government officials were poor. Xi's job is harder; he has to say, "It's OK for most people to not get rich." It is especially important to the government for young people to internalize the Dream; the instructions to educational institutions to begin promoting Xi's formulation were leaked from the General Office of the Central Committee, who is responsible for drafting and circulating party directives.

The threat the Chinese Dream is meant to counter is the creation of a public sphere, a place where the public can discuss the gap between how things are and how the public would like them to be. In Benedict Anderson's brilliant book *Imagined Communities*, he describes how local newspapers became the place where colonized Indonesians formulated an identity separate from, and ultimately in opposition to, their Dutch rulers. The Chinese government has had tight control over media for the last two generations; the spread of the internet marks the first time the Chinese have had anything that works like a public sphere, a change that has become a significant preoccupation of the current government. Various state mouthpieces began discussing Chinese public opinion on the internet in almost apocalyptic terms. *People's Daily*, the most important state newspaper, ran an editorial saying that online media "have already become the chief battleground in the public opinion struggle, and their importance and status in the overall news and propaganda framework is ever more obvious." A year later, the People's Liberation Army paper published one saying, "The Internet has grown into an ideological battlefield, and whoever controls the tool will win the war."

Chinese social media still has lots of celebrity gossip and trading of funny images, of course, but there is also real political conversation, so much so that the government is having to crack down on political speech not just by academics and media outlets but by citizens. In 2013 several Weibo users with audiences in the millions (the so-called Big V users, whose identities have been officially verified) were arrested or intimidated for talking politics. This helped remove some of the overtly political speech from the public sphere, but as the government is discovering, as a country gets a middle class, their demands change, and the definition of what constitutes politics changes with it. Politics now includes issues of food safety, building construction codes, transportation safety, and so on. Some of this material is useful to the government—in the aftermath of a horrific high-speed train crash, the government allowed criticism of the railway minister to flourish online, as a precursor to sacking him—but as the government has seen over and over, any widely available forum for public complaint turns to demands to root out corruption in the government.

Like all authoritarian countries, China is terrified of social media. Some of this is fear of media's synchronizing power in general—German sociologist and philosopher Jürgen Habermas pointed out in his brilliant but almost unreadable book *Structural Transformation of the Public Sphere*, political argument playing out in public media is bad news for elites. This has particular resonance for China, which looked to the Soviet Union for instruction, first for guidance and then as a cautionary tale. Ever since Khrushchev denounced Stalin and demoralized the U.S.S.R.'s leadership, it has been an article of faith among the Chinese Communist Party that ongoing disagreements between leaders are not to be aired in public, leading to the spectacle of officials assumed to be in good standing right up to the day they are arrested.

This reflexive fear of social media has latterly been accompanied by practical examples. The ouster of Joseph Estrada in the Philippines was largely coordinated by SMS (and, ominiously for China, the thing that turned Filipinos out in the streets was anger over corruption). The Color Revolutions in Ukraine and Moldova, former Soviet territories, saw citizens trying to escape Russia's sphere of influence and join the E.U.—a people rising up to demand an escape from the conditions of autocracy is bad news for anyplace that styles itself a People's Republic.

Given that rich populations are more demanding, and that political expectations are expanding to include quality of life issues, censorship is no longer adequate to keep the incipient public sphere under control. This is where propaganda comes in. Xi first mentioned the Chinese Dream as a signature of his coming reign back in 2012, the year he took over. He doubled down once in office, insisting that the Dream "deeply reflects the Chinese people's dream today and is consistent with our glorious tradition," a sentiment that lets him take credit for progress and tradition in the same breath. Shortly thereafter, posters lauding this Dream appeared all over China; the most recognizable symbol is an apple-cheeked girl in red silk, her plump face resting on her hands as she gazes dreamily at the viewer.

The posters come with captions like "From morning to evening, approaching the dream," hardly a call to any particular sort of action. The non-economic content of the Chinese Dream is a hodge-podge. There is genuflection to Confucius (whom Mao loathed), emphasis on ancient mistreatment by enemies everywhere, references to 5,000 years of Chinese tradition, and to modern China. The Dynasties were glorious, Mao was glorious, the undoing of Mao's policies were glorious, and so on. In urban areas especially, billboards touting the Dream are everywhere, in an attempt to create a sense of social solidarity, without committing its rulers to any particular course of action.

No one knows whether the Chinese Dream will improve social cohesion in the long haul. Even given the general difficulty of political prediction, the future of China is a hard case. The problem isn't just that no one knows what will happen in the "events, dear boy" sense; the future always includes surprises. The issue with predicting China's future is more fundamental. Even if China has a completely non-surprising decade between now and 2025, there is no way to predict how things will go. China's economic growth is singular; no country in the world has ever achieved what China has achieved in the last forty years. And their political goals are as well; no country in the world has ever achieved what China intends to achieve in the next ten years.

In the history of industrialized states, many single-party systems have lasted into middle age: the Liberal Democratic

Party in Japan, the Institutional Revolutionary Party in Mexico, the National Democratic Party in Egypt, the dictators who grew old in office—Castro, Tsedenbal, Hoxha, Bongo—and, *primus inter pares*, the Communist Party of the Soviet Union. None of them made it to their seventy-fifth birthdays still in charge. If the Chinese Communist Party remains in undisputed power over the next ten years, it will be the first modern single-party state ever to do so. There is no natural law that limits authoritarian regimes to three-score and ten, but there is also not one example of a single-party state lasting that long. No matter what happens in the next ten years, a whole host of perfectly reasonable assumptions are going to be shredded one way or another.

One small ramification of this uncertainty is a need for less hubris among American foreign policy observers (a habit that might be profitably expanded to other regions). A far more important one, however, is that the Chinese government has no idea what will happen either. They have been keen students of anti-authoritarian uprisings; the end of the Soviet Union, the Color Revolutions ten years later, and the Arab Spring all provide examples of seemingly stable governments that collapsed in months (and sometimes weeks). In 2013, Xi commissioned a new study of the collapse of the Soviet Union, a worry and theme in Chinese communist circles since the end of Stalin's rule. The theory he seems to favor is that the Soviet Union collapsed because of incompetent leadership, not systemic weakness.

Understanding that uncertainty provides a way to make sense of many of the party's other actions. Under Xi, there is a systematic attempt to rein in any possible source of coordination that is alternate to the government—not even opposed to,

102   but simply alternate to. Trying to restrain the use of puns or time-travel, shutting down film festivals not because they hate film but because they fear public gatherings, arresting feminists and moving non-governmental organizations under supervision by the Public Security Bureau—all of these things make it clear that their greatest fear is the risk of coordinated refusal, by their own people, to accept the Chinese Dream on offer, and ask for a new one.

Copy the Copycat

Though China has had the "Open for Business!" shingle hanging out for some time now, even today China is tremendously isolated, by virtue of travel, ethnicity, and language.

To say China is isolated by travel seems a stretch; it is the fourth largest site of inbound tourism in the world, after Spain, France, and the U.S. But everything in China has to be measured against the denominator of its population. In 2013, Spain received 150 tourists for every 100 Spanish citizens, half again as many tourists as there are people who live there. France received 120 tourists per 100. And China? Four. Four tourists a year per 100 Chinese, and a sizable chunk were from Taiwan or Hong Kong—Chinese who have to use their passports to get into mainland China.

Outside international neighborhoods in just a few enormous cities—especially Beijing, Shanghai, and Shenzhen—seeing foreign faces is still an event. I live in the largest, most cosmopolitan city in China, but my apartment is in a compound where there are almost no other foreigners, and even now merely walking through my neighborhood I can turn heads. When my children were younger, people would come up on the street and pat them on the head, apropos of nothing. The kids hated it, but it was never done out of anything but curiosity and affection. This interacts with the second source of Chinese isolation, which is ethnicity. The government makes much of the country's dozens of ethnic groups—6 million Mongols, 10 million Uighurs, and so on. But there are 1.2 billion Han Chinese out of 1.3 billion total residents. China is remarkably, overwhelmingly homogenous, 92 percent Han, an imbalance that is even greater in the urbanized east than the 11 to 1 ratio would suggest. Among older people and in smaller towns, not only are there still those who have never interacted with anyone from a different country, they have never interacted with anyone of a different ethnicity.

The combined sense of solidarity and difference make China easier to govern, because people internalize a sense of specialness that can be used to explain why they aren't allowed to vote for national leaders, or read the foreign press. In the aftermath of any real faith in communism, government gestures toward solidarity almost always involve a sort of ethnic nationalism, with its constant emphasis on "5,000 years of Chinese history," a phrase too often repeated to be anything but a coordinated effort. The actual concept is meaningless—China has been a dynasty, a republic, a failed state, a dictatorship, and a technocracy in the last hundred years alone. Reaching further back, there are centuries of dynasties alternating with warring states and warlord rule, and constantly shifting borders. There

is no political or cultural continuity that started in 3,000 BCE.
The common list of dynastic succession that schoolchildren
learn leaves out many periods where there was no one in particular in charge, but lots of groups who thought they should
be. Despite the historical inaccuracy, however, claiming vast,
unbroken cultural continuity is politically useful.

The characteristics of homogeneity and isolation have been
prized assets for Chinese governments for centuries; the tension the government now faces is how to continue to get the
economic value of opening up while preserving the political
value of closedness. This problem is general in East Asia. Even
in autocratic governments like Laos and Vietnam, the shift to
market economies is widespread. Myanmar, until recently
almost as closed as North Korea, has announced that it is
open(ing) for business. For Americans, it is striking to see that
even in Vietnam, the only communist country to defeat the U.S.
in a war, democracy lost but capitalism won. That is the case in
China too—without a shot being fired, the communist economy has been replaced with a capitalist one. (The government
prefers "Socialism with Chinese characteristics" for the obvious reason.)

The Chinese perfected open markets but closed government even in defeat. After losing to the British in the Opium
Wars, the Qing Dynasty successfully negotiated a peace treaty
that largely limited foreigners' access to a few "treaty ports,"
principally Hong Kong and Shanghai. During the Cold War,
they made it illegal to make or promulgate accurate maps of the
country (a situation that persists to this day in the intentional
inaccuracy of GPS maps on consumer devices). Similarly, they

have to manage both sides of the linguistic barrier between Chinese and English. Though it's not widely appreciated in the U.S., China itself is one of the principal exporters of English, because English solves the same problem for China as it does for Europe. In an area of linguistic diversity and historic enmities, there is no way that the Chinese could or would translate street signs into Thai, Korean, and Japanese, much less Hindi, Mongolian, and Tagalog. As long as everyone speaks the same second language, the Chinese can translate things into one language and reach the largest possible market.

This has famously led to an appreciation of English interpretations that sound odd to native speakers. (A sign at our local swimming pool reads, "Non business hours, please do not into, therefore all the consequences." Not the King's English, but it does the trick.) It also leads to an appreciation of the roman alphabet as a sign of class, even when these words are used more as sigils than expressive encodings, as with my favorite local firm, a cosmetics company with the evocative but unpronounceable name FANCL, or one of my family's prized possessions, a guide to our rice cooker whose cover admonishes, "Read all instructions carefully"—the only English words to appear in the entire manual. Signs push English-language instruction for everyone from businesspeople ("Wall Street English") to children ("American Baby International English"). Despite both needing and loving English, there is a simultaneous campaign to limit its use.

One of the early signs that Xi Jinping's media crackdown was entering new and more serious territory was the sudden shuttering last fall of every Chinese site offering amateur subtitling of English movies and TV. The effect of this coordinated purge was not so much about limiting the flow of subtitled movies—the Chinese market for pirated DVDs is famous the world over—as about limiting the threat of widespread or synchronized viewing of Western content. Unlike internet video, DVDs and movies in theaters both cost money, which dampens their spread. DVDs also can't be shared on social media, but URLs can, meaning that a piece of video could be viewed by millions of people without much government oversight. The Chinese go further to segment the market for subversive thought by ability to pay than any government in the world. The general strategy has been to use every possible barrier—natural and manmade, linguistic and cultural—to ensure that outsiders operate at a disadvantage.

You can see this pattern in the government's recent crackdown on VPNs. A virtual private network is a way of hiding internet messages from observation by the various computers those messages pass through between your computer and your destination. From the user's point of view, a VPN is a product, a piece of software that sits on your computer or phone (a silly distinction, but widespread). You turn it on, you can get to blocked resources, you turn it off, and you're back to seeing the internet the Cyberspace Administration of China wants you to see.

A VPN uses an intermediary service to hide and then redirect your traffic; the effect would be a bit like hiding the destination of a letter from the U.S. Postal Service by writing and addressing a letter, and then putting that in a second letter to a friend in another country, who would open your mail to them, take out the original letter, and mail it on to its destination. Similarly, a VPN service provider can receive a message from your computer, forward it on to a third party, and then send the result back to you. So if you are behind the GFW and want to use Google (or watch a video on YouTube, or read Twitter), you connect your computer to a VPN server in South Korea say, or Australia, and send that computer a message that says, "I want to see this YouTube video." The VPN server then contacts YouTube on your behalf, and sends the message back to your computer, encrypted in such a way that no one but you knows its contents. Through this indirection, you can punch a (virtual) hole through the Great Firewall.

China was in the news in early 2015 for cutting off access to VPNs. Except that the story, as reported, was wrong. China didn't cut off access to VPNs, it cut off some access to some VPNs, and that difference makes all the difference. In the Chinese context, anyone offering VPN services is a company that, by definition, specializes in punching holes in the Great Firewall. It's easy to imagine that the Chinese government might take this amiss. It's harder to understand their actual attitude toward VPN providers; since all VPNs route traffic to well-known destinations outside the country, China could cut off VPN traffic at a moment's notice. They don't, because despite being forbidden, China needs VPNs, or, put more specifically, the government needs some classes of employees to have access to the uncensored internet. For forty years now, the profitable exchange of ideas and deals with the rest of the world has been essential to the Chinese economy, and the government can't afford to cut off that air hose. However, it also can't afford to let everyone have access to it either. They treat open access to the internet like oxygen, necessary in small amounts, but fatal if either too prevalent or absent.

VPNs are ubiquitous in rich, educated China, and will never go away as long as the firewall exists. Businesspeople use them to get access to services like Gmail and Facebook; kids use them to get access to games and porn. A VPN is like a Special Economic Zone for the internet, a small part of China carved out for increased freedom and reduced oversight. And, like the SEZs, the Chinese worry is how to keep them running without allowing them to spread. Unlike an SEZ, however, VPNs don't stay put. Electronics and software get cheaper and easier to use every year, so any digital boundary that gets established between the elite and masses is likely to change in the direction of mass access. This happened with VPNs. Though much of middle-class China is perfectly happy with China as an intranet, a protected network intentionally disconnected from much of the rest of the world, the growing number of Chinese who travel to or study in countries outside China, or who have business there, want unrestricted access; VPN providers, in the business of collecting a few dollars a month to help provide that access, are happy to oblige.

This use of VPNs doesn't carry the feel of protest; China has a long tradition of citizens jumping over barriers of all sorts, from lines for movie tickets to traffic cones, and the firewall is just another one of these. I've talked to many people who get past the firewall for one reason or another, from businesspeople to employees of internet firms, and the consensus is acceptance that the government is within its rights to control internet access in general, accompanied by a sense of pique that their particular needs are caught up in that control. ("Censorship OK", said one businessperson, "but don't block Gmail! I need that for work!") So when China blocked VPNs, the ones it targeted were the most public, the cheapest, and the most used on phones (where the public threat is). The VPNs they left alone were less widely used, expensive, and mostly used on PCs (where the business need is).

And of course the talk around myriad dinner tables the evening of the blockade was which VPNs still worked. (I know someone who subscribes to five different VPNs, to increase robustness against the targeting of any particular one.) This may seem to defeat the point of the blockade, but the Chinese, as ever, are not trying to discomfit bilingual citizens or coastal elites. They just don't want the general public to enjoy the same degree of freedom those elites do. By targeting the most public VPNs, they created a huge disincentive for any of us who knew which ones were still working to say anything about that out loud. For me to go on WeChat and say, "Hey everybody, Herpaderp VPN is still running" would get Herpaderp killed pretty fast. And so after the crackdown, the information about what was still working did not get a public airing—if you are cut off, you need to know someone who knows how to reconnect. People with those connections are likelier to get what they want, but less likely to complain in public, precisely because of the skein of mutual obligation.

Elites have to have access to the outside world, both in order to enable import and export businesses and to keep them happy. A surprising number of children of privilege go to college in the

U.S. (including Xi Jinping's daughter, recently graduated from Harvard). Given the relative ease of getting U.S. work permits, brain drain is a constant, low-level risk. Meanwhile the majority of the country can't be allowed that same sort of access. The government wants to keep ordinary citizens from communicating with the outside world (and especially with ethnic Chinese who have grown up in democracies), and to keep them from using tools outside government control, like Facebook, Twitter, and Meetup, to coordinate with one another. Making life better for elites than the masses is what most societies do, practically by definition, but the Chinese case is complicated by their insistence on extending this dual system to media, and especially to electronic media.

As Xi's government continues to tighten its hold on the media environment while electronics become more capable and cheaper, it is becoming progressively harder to maintain the desired equilibrium between elites, where access is tolerated, and the general public, where it is feared. Xiaomi exemplifies this problem. Being big in the Chinese market is a great advantage, of course, but the logic of a globalizing Chinese business is that while the Chinese market is bigger than anybody else's, it is not bigger than everybody else's. Their global goals mean offering services elsewhere that are disabled for their home market. Every phone they ship can be configured to support a VPN natively. They run their own Facebook page, Twitter feed, and YouTube channel, all blocked in China. They apologize for not installing Google apps. At state-owned enterprises, employees are judged on a degree of fealty to Chinese Communist Party rhetoric. At a firm like

Xiaomi, in the business of giving people what they want, censorship is nothing more than the cost of doing business. It is always in Xiaomi's interest to sell users the most freedom the government will allow.

At the same time, Xiaomi has to signal its desire to be the kind of firm the government expects it to be. Xiaomi made the Chinese news in the middle of 2015 for opening its first Communist Party committee, which provides a more direct route for the government to both understand and instruct Xiaomi's strategy. These committees are required for state-owned enterprises, like banks and phone companies, and for state-run organizations, like hospitals and schools. They are theoretically required in private firms that have more than three party members as employees, but the enforcement of this rule is somewhat lax (as Xiaomi's years without a committee demonstrate). If a firm is either rich or runs a service the government sees as vital, having a party committee starts to look a lot less optional.

The arrival of the committee is another marker of Xiaomi's size and importance—internet giants like Baidu and Tencent also have them—but the closer the company's ties with the Chinese government are, the more obvious the political dimensions of mobile phones become. Some Chinese citizens took to Weibo to vent disdain for the move. One asked, "So here is the question, [shall Xiaomi employees] listen to the Party Secretary or Lei Jun?" Another said, "Hehe, so the Chinese technology companies can never be GOOGLE [spelled in English], Wikipedia, those companies which are so welcomed and respected by the world."

One particularly elaborate Weibo message was filled with the sorts of puns for which Chinese internet use is famous: the simple translation is "Xiaomi founded the party committee? And I heard that it is called: Mi Gong?" but the poster substituted a homonym for "party committee" meaning "crotch," and *Mi Gong* is an elaborate play on Xiaomi's product naming. The poster took the character *mi* from Xiaomi and *gong* from *gong chan dang,* the Communist Party, but the characters for *mi* and *gong* can be combined to make the word "shit." (Unlike the shorter posts above, this one has since been deleted.)

The party representation can also create friction with trading partners. In 2009, a U.S. House of Representatives committee investigated Huawei, the Chinese electronics firm. Their final report included alarmist language about the party committee: "Huawei admits that the Chinese Communist Party maintains a Party Committee within the company, but it failed to explain what that Committee does on behalf of the Party or which individuals compose the Committee." Some of the "communists under the bed" tone is naive—having a party committee serves some of the same purposes in China as hiring a lobbyist does in the U.S.—but the effect on Huawei's ability to sell its products in U.S. markets is affected all the same. The U.S., like China, is happy to use national security arguments as trade policy implements. Xiaomi doesn't want this kind of friction, either from Chinese users or foreign ones, but as its importance grows, and the government's concern about social media grows, it doesn't have much choice.

Given the Chinese government's historic enmity and current need for communications tools among its citizens, what does Xiaomi mean for this balancing act? In business terms, it is an early, wildly successful example of a global, design-centric, service-oriented Chinese company. For most of its 100 million customers, this doesn't mean much more than access to good cheap products. For the subset of folks who are Mi Fans, Xiaomi stands for a kind of youthful energy, a success story that is not just about cheap assembly. For its investors, Xiaomi is a company whose valuation has appreciated something like 18,000 percent since its initial funding, not bad for a few years' work. Told this way, Xiaomi is just another improbable digital success story, another example, alongside Apple and Alibaba and Amazon and Tencent, of "how we do business today."

Xiaomi also means something for how the world will get connected. Mobile phones are the most broadly desired category of complex goods in the world, beating out their only rivals, cars and televisions, by a country kilometer. The mobile phone is also becoming the universal source of connectivity for most of the world's population, increasingly the gateway to every form of communications other than face-to-face, to every form of content other than paintings, and to every form of commerce other than haggling. Thanks to the mobile phone, the developing world, and therefore a majority of the human population, has gotten connected in the last twenty years. In the next ten, a majority of them will move from simple phones to real networked computers. Though Apple invented the smartphone, and Samsung spread it, it is Xiaomi who showed the world how to create a defensible market between luxurious and crappy, and to scale up to meet the

rising demand of the rapidly expanding and increasingly global
middle class.

Even if their incredible early start fades, and Xiaomi eventually becomes just another electronics manufacturer, its legacy, already secure at five years old, is this: They have shown everyone else how it's done. Xiaomi is what a twenty-first-century manufacturing company looks like, an early example of a type of company that we will all get used to over the next few years. By showing both its customers and its competitors that consistent attention to users and constant software upgrades is a good customer-loyalty strategy, Xiaomi has helped shift competition from hardware, which customers upgrade once every few years, to software, which they can upgrade once every few days. The signal achievement of digital devices is that software can teach old devices new tricks. That achievement is moving from the PC and the laptop to the phone, and therefore to almost everyone in the world, by 2025. However, it's not a revolution if nobody loses, and in the model of twenty-first-century manufacturing—software-driven, service-led, sold online—there are plenty of losers.

In the category in which Xiaomi competes most fiercely—Android-based smartphones—Apple has already lost ground. Apple doesn't need anyone's pity, of course. It is, by many measures, the most successful company in the world. But of the two dreams that company pursued—to be a wildly successful business and to be a world-changing organization—the latter turns out to happen mainly through copies. Just as Microsoft made the graphic user interface ubiquitous by copying Apple, Apple brought the smartphone, touchscreen, and app store into existence, but most people will experience those things courtesy Apple's competition. Android already has a billion users, something the iPhone may never have.

Those economics are getting more adverse—between 2010 and 2014, the average price for an iPhone without a contract fell from $702 to $657, a decrease of a little over 6 percent. In the same period, the average no-contract Android phone fell from $441 to $254, a decline of 42 percent, leaving Android phones costing, on average, two-fifths of what an iPhone costs. Imitation is the sincerest form of flattery, but damaging to the revenues of the flatteree. The cheap end of the market was never going to be Apple's, but Xiaomi's ability to deliver high quality and a moderate price tag has shown other firms selling Android-based phones a pattern they can copy. This competition—better hardware and software at lower prices— makes reaching the middle of the market increasingly difficult for Apple as well.

At least luxury comes with high margins. Xiaomi is worse news for Samsung, the previous vendor at the middle of the cost-quality curve and the largest phone manufacturer in the world, but one with a shrinking market share in the world's biggest market. Samsung's market share worldwide fell by a third in a single year, from 30 percent in the last quarter of 2013 to 20 percent in the last quarter of 2014. Of course, this wasn't all the doing of Xiaomi, whose move worldwide has just begun. Samsung is instead being challenged by the category of manufacturer generally shown on analyst charts as "everybody besides Samsung, Nokia, and Apple," the cumulative force of the dozens of smaller manufacturers. Xiaomi is now the leading company in that group, and the risk to Samsung is that enough of those other companies learn from Xiaomi that the overall competition gets even harder.

And, in the spirit of revolutions eating their young, Xiaomi may well become a victim of its own success as well, if enough of those "everyone else" manufacturers copy it. The company has a bright future, because a good, cheap smartphone is one of the most widely desired products in the world. But in the same way that Apple opened up the market for Android phones, Xiaomi is teaching other businesses how to compete in the market they helped create. Xiaomi's public appeal and its reliance on online sales to control both marketing and sales costs are not hard acts to copy. As the world grows richer, and a smartphone goes from being optional to being essential for nearly everyone, there are a handful of emerging markets—India, Pakistan, Malaysia, Indonesia, Nigeria, Kenya, Brazil, and Mexico—that will account for a huge proportion of sales. As a global design firm, Xiaomi will either give these users what they want, or will turn out to have shown its competitors how to do so.

After Xiaomi, it's possible to see that a small group of passionate users can provide huge value for design, debugging, and proselytizing, and that good design for mid-priced products creates significant value in a crowded market. Competitors are already attacking Xiaomi using the strategies Xiaomi perfected. Established brands like Huawei and ZTE in China and HTC in Taiwan are moving to improve design and customer service, and emphasizing their online sales. OnePlus, another Chinese brand with global ambitions, is competing on hardware. Its second phone on the market, the amusingly branded OnePlus 2, costs about the same as a Mi4, but with more storage and a bigger screen. Other companies, like Meizu and Coolpad, are competing on price. Meizu recently copied Xiaomi with an own-brand version of Android, the oddly named Flyme OS. (Shades of FANCL.) Meizu phones are also trying to clear a space for themselves by selling nice design at a lower price tag, except their competition is not Samsung, but Xiaomi. Meizu recently got funding from Alibaba, China's main ecommerce platform, where Xiaomi has previously dominated Singles Day sales. (Alibaba has also invested in Xiaomi. With Chinese startups, there are no conflicts of interest, only ironies.) Coolpad, a 20-year-old electronics firm in Shenzhen that has mainly made hardware for other companies to brand and sell (an "original equipment manufacturer," or OEM), recently launched a consumer brand, Dazen, available in cheap and very cheap, and only sold online. And Meizu and Coolpad have both launched in India.

The nationalist appeal of Xiaomi as a local champion is also proving easy to copy. Indian firms are now reacting to Chinese competition. Micromax, headquartered in Haryana, has launched a new brand, Yu, and a phone, the Yuphoria, which uses a stripped-down version of Android called CyanogenMod. Like Xiaomi, they sell online only, albeit through Amazon, and they use their customer forums to tout rapid iterations of their software. There are more from where the Yu came: India also has Celkon, Obi, and Wickedleak, purveyors of the Wammy phone. (The hardware for these firms, of course, is sourced from Shenzhen.) This pattern is also spreading to developed-world markets. Wiko, a French phone company, went from concept to company when the founders were shopping for parts in

Shenzhen (as one does). Wiko had trouble raising money—few investors believed a new European phone company could succeed—so they took an investment from the Chinese manufacturer Tinno Mobile. Wiko is thus mostly Chinese, both owned and supplied by Tinno, but given its thin veneer of French design and marketing it looks like a local firm to the French. The resulting excitement over Wiko as a homegrown business helped them to become the second largest phone vendor in France (after Samsung, as usual). This preserves the pattern of "designed elsewhere, made in China," but with the twist that ownership, not just sourcing and manufacturing, has now moved to China as well.

Competition this fierce can't last, of course. GSMArena.com currently lists 103 mobile phone manufacturers; you likely do not know anyone who's bought a phone from almost any of them. (The BLU Studio Energy, the Parla Zum Bianco, the Yezz Billy 4.7, the Eten glofiish [sic].) Given the prevalence of the "black slab of glass" design and the touchscreen as the universal interface, there are far too many phone companies in the world today to survive. In a reversal of the classic logic of business disruption, mobile phones are one of the few markets where the good products can get cheap faster than the cheap products can get good. The next five years are going to see a wave of mergers and bankruptcies in the phone business, and for any manufacturer not named Apple, copying at least some of what Xiaomi has done will become imperative to survival.

Moving away from the narrow market for phones, Xiaomi is bad news for the models of manufacturing that start with hardware first, rely on unit sales without engaging users, and sell things in stores. There will always be some products that have to be made and sold that way, of course, but everything with a chip in it (an increasingly sizable subset of everything) is a good candidate to be sold services-first, and online. As we know from companies like Tesla, even cars, the most valuable manufactured item anyone buys, can be sold online, with only protectionist impulses keeping local dealerships free from competition. China again has a leapfrogging advantage here—because all of its commercial infrastructure is recent, new ways of doing things get metabolized faster in China. The car company Daimler decided to try a flash sale of their two-passenger city car on WeChat and sold 388 of them in three minutes. Imagine them trying that on Facebook.

By demonstrating that a Chinese firm can link good design with global ambitions, Xiaomi also portends a gradual end to the "designed elsewhere, made here" pattern. Xiaomi's incredible financial success is large enough to cause people to reconsider what they had long believed about manufacturing companies. By starting with customers before they had a product, Xiaomi managed to get design advice and free marketing as they grew. By starting as a services company that happened to make hardware, they were able to design a hardware pipeline that kept the complexity of their product range to manageable levels. By going for frequent low-cost upgrades in software, they can keep a given hardware configuration on the market far longer. (This may seem like a niche realization, but your refrigerator, TV, and car all have software, and it's coming to your coat, your desk, and your bed.) Xiaomi has invested in Midea, a Chinese appliance firm, and is creating new designs for that company's existing products, starting with an air purifier. It is applying its expansion model to other products, from fitness trackers to sports cameras to big-screen TVs to their upcoming drone. All of these items can be improved by software upgrades.

Xiaomi's sales model will be copied as well. The rolling Götterdämmerung that is online selling—which came first for travel agents, then record and video stores, then electronics, books, and games—is now headed for every form of retail other than manicures and burger joints. Retail sales is one of the last no-degree-required, train-on-the-job occupations left. Its automation leaves a world where you would only sell things in stores as a forced move (This Yogurt Can't Be Sold Online) or as a shopping experience, not just a simple service (Buy Kale Here and Feel Sanctimonious). Xiaomi didn't invent the pattern of online-only sales, but by selling and shipping very expensive hardware that way, and shifting the cost and risk of physical retail to their partners, they make the advantages obvious. If you were starting a company tomorrow, you'd only sell it in your own stores if you had to.

Xiaomi's model could work anyplace there is a budget market for design, and where the resulting products could be networked. Though the so-called Internet of Things is still mostly in the future, it is already clear that one of the main business models will be Xiaomi's: Any device with a network connection can generate enough revenue from services to offset the cost of the hardware. Their famous involvement of the first hundred users, and their constant soliciting of feedback, move the pattern of lead user innovation from edge-cases in open source software and athletic gear to feature suggestions

122   and improvements for one of the most complex and widely desired products in the world. Xiaomi marks the end of "lead user innovation" as an interesting edge-case, and its arrival in the mainstream.

For China itself, Xiaomi is both opportunity and challenge. As proof that the Chinese entrepreneurial class can now compete on design, service, and customer satisfaction, it is a huge boost for the country's place on the world's economic stage. But as a firm that has to make two versions of its software, with the one designed for its home market intentionally giving its customers less freedom, while running into patent trouble, licensing issues, and governmental suspicion abroad, it is also an example of the growing risks to the strategy of practical isolation that China has always practiced. You can get a sense of the patent-pool imbalance from the numbers: Xiaomi filed for around 1,000 patents related to mobile phones in 2014. Ericsson holds something like 35,000, many of which cover far more fundamental aspects of connecting a phone to a network. There is also a potential suit against Xiaomi for not releasing some of the source code for its products protected under open source licenses. It will be easy to sue for this behavior in American courts, should Xiaomi ever sell phones in the U.S. China's relatively protective regime for local companies facing foreign patent challenges is not portable outside its borders.

Predicting the future of China is a mug's game. The problem isn't just that by three o'clock on any given afternoon, someone will have published an essay entitled "China's Imminent Collapse" and someone else will have published "China: Superpower of the Asian Century." The problem is that both

essays will be plausible. China today is one of very few political
entities with the resources and government to shape the world
as a side-effect of its actions. It also faces a set of problems that
no single-party system in modern history has ever successfully
kept at bay as long as China will need to, if the current system is
to persist for even ten more years. Using the market to gradually
fix a totalitarian government is like making a pot of tea by running a volcano through a glacier. It's possible to get the balance
just right, but the forces of conservatism and corruption always
threaten to freeze progress, while an economy growing this fast
is often at risk of overheating. Both threats have happened in
the last forty years, but the country has always re-balanced in
time to keep the party in power and economic growth on track.

What we can predict—or at least notice—is when the
underlying logic of the situation changes. We can see the end
of China as the world's workshop and its emergence as an orig-
inator of new products for a global market—something I first
understood when I accidentally bought a Mi3 and became sud-
denly and briefly cool. One source of China's success in keeping its open economy and closed political system running on
separate tracks was the near-total clarity of the "you design, we
replicate" pattern. It will be relatively easy to escape that model
for firms that make non-political goods. Chinese companies
like Haier and Galanz aspire to become global brands for appliances. But it will be harder for Xiaomi, which occupies an almost
unique position between offering simple hardware and complex
services as a bundle. China demands of all service companies
that they not give consumers capabilities that the government
doesn't want its citizens to have. That strategy in turn works

better when Chinese citizens can't see what consumers in the rest of the world have access to. But as Xiaomi and its inevitable copycats make it increasingly clear that they are happy and even eager to provide the freedoms users want for anyone who buys their products off Chinese soil, the story of the company's success will become harder to tell in the official context of the Chinese Dream.

There isn't a power struggle between the government and Xiaomi. Indeed, Xiaomi's successes are evidence of the government's accomplishments in maintaining a dynamic economy. There is, however, a tension in their respective goals. Phone companies sell freedom, and the configurability of the smartphone doubles down on that promise. The carrot of the Chinese market is huge, of course, but for a company with global ambitions, that may not be enough. China is bigger than anybody, but it is not bigger than everybody, and tensions from operating on both sides of those national borders are growing. Despite the firewall, Chinese companies continue to advertise themselves on Facebook and Google—my dentist in Shanghai puts his Gmail address in scrolling LEDs in front of his practice. To do business with the rest of the world, Chinese firms increasingly have to get good at using services that are both essential and (theoretically) unavailable.

As this shift goes on, we do not know whether the Chinese Dream will keep its citizens from demanding things the government won't—or can't—deliver. The Chinese Communist Party long ago stopped claiming that they have discovered The Way (or even A Way) relevant to all the peoples of the world. (The Shining Path in Peru, rather than the CCP, is the last significant

Maoist organization in the world.) With the desuetude of Mao Zedong Thought, and having long abandoned what zeal they had for positing that communism provides any sort of universal ideal for arranging a society, the central source of legitimation for the Chinese government in the eyes of its people has been "How 'bout that GDP?" As that period ends—and it is ending now—the balancing act China is committed to is as complex as anything they've tried since the 1970s.

A century is an invented unit, as manmade as a mobile phone, as artificial as a Big Mac, and yet it's a useful marker for all that. We don't know if our great-grandchildren will regard this one as the Asian Century, but it will be far more Asian than the twentieth century was, when poverty and isolation kept Asian countries and especially China largely sidelined from global influence. That period is ending, and you can see its end in China's struggle to balance opening up and closing down, as its most successful new firms turn global. We don't know what the result of all this new connectivity will be, for China or for the world, but we can at least say, "We don't know when the old model ended, but we can now see that it's gone."

We are living in a golden age of writing about China. A generation of culturally fluent observers, living in an increasingly open China, is transforming the old art of "China watching" into much more engaged reporting, often by people who are participating in Chinese life. These recommendations start with contemporary China, working backwards.

Evan Osnos's *Age of Ambition: Chasing Fortune, Truth, and Faith in the New China* (FSG, 2014) chronicles the disorienting changes in China's society when ordinary people began to aspire to something more than going along and getting along. The rise of ambitions outside the country's historical elite, accompanied by massive demographic and economic change, altered many Chinese norms, and Osnos captures both the excitement and the disruption of these changes beautifully. This is simply the best book on China today.

Peter Hessler's *Oracle Bones: A Journey Between China's Past and Present* (Harper, 2006) is an earlier look at similar territory. Hessler's book, researched and written a decade before Osnos's, looks at the accelerating changes in a country that had emerged from decades of political and economic hardship into calm, then stability, then growth. Hessler's *Country Driving: A Journey Through China from Farm to Factory* (Harper, 2010) is also marvelous, providing a fascinating citizen's-eye look at the country's transition from rural to urbanized society.

James Fallows, *The Atlantic*'s China correspondent in the first half of the 2000s, collected his observations for that magazine in *Postcards from Tomorrow Square: Reports from China* (Vintage, 2008), a title that refers metaphorically to China's future and to Fallows's real-world residence in Shanghai's Tomorrow Square. (The chapter "China Makes, the World Takes" was an early look into the manufacturing juggernaut of the industrialized Pearl River Delta.) Fallows is also the author of *China Airborne* (Pantheon, 2012), which looks at the economic and social changes in China through the lens of air travel, both as an economic and social part of Chinese life.

Many of the changes that created contemporary China were unleashed by Deng Xiaoping, the leader most responsible for dismantling Mao's China. Ezra Vogel's *Deng Xiaoping and the Transformation of China* (Belknap Press, 2011) is the best guide to Deng and that transformation after Mao's death. (The book is marred, in the last chapters, by downplaying the end of the 1989 Tiananmen Square uprising as a "tragedy" rather than a massacre.)

The previous books concern themselves with the incredible changes in China after Mao, but no era ever starts with a clean slate. Both Deng and Mao inherited many cultural and political characteristics shaped far earlier. Jonathan Spence's *The Search for Modern China* (Norton, 1990) traces the development of China into a modern bureaucratic state, starting in the seventeenth century (roughly the same time as northwestern Europe's modernization). The book ends with the early Deng years, but its great strength is how the culture and politics in the eighteenth and nineteenth centuries affected China in the tumultuous first half of the twentieth.

And if you want a survey of the sweep of Chinese history, there is the ten-part online class from Harvard, called simply *China*. Put together by professors Peter Bol and William Kirby, and narrated by Christopher Lydon, *China* will let you start from pre-history and work your way through to today, or dip into the periods you are interested in: www.edx.org/xseries/chinax.

The End of Copycat China: The Rise of Creativity, Innovation, and Individualism in Asia (Wiley, 2014), by Shaun Rein, documents the changes in the Chinese business environment, from manufacturing products designed elsewhere to creating new products locally. It follows business-book conventions (it has key action items for the reader), but the chapter on local adaptations to rising pollution alone is worth the price of admission.

China's Disruptors: How Alibaba, Xiaomi, Tencent, and Other Companies are Changing the Rules of Business (Portfolio, 2015), by the consultant Edward Tse, covers a similar theme, documenting the rise of globally competitive firms like Xiaomi and Haier. Tse is less skeptical about the ability of China's internet firms to expand outside the Chinese market than I am, but the book is a good guide to entrepreneurial success in China.

China's unprecedented media explosion—rapid, large-scale adoption by the people and the subsequent attempts at surveillance and control by the party in response—is the subject of *Changing Media, Changing China* (Oxford University Press, 2010), by Susan Shirk (no "y" and no relation). Shirk documents the transition from a China with no journalism, only propaganda, to one with a lively but hotly contested public sphere.

Emily Parker also wrote a fascinating book on political dissidence in authoritarian regimes called *Now I Know Who My Comrades Are: Voices from the Internet Underground* (Sarah Crichton Books, 2014). The book covers three countries—China, Cuba, and Russia—but the section on China is the leading example, and concentrates on the pro-democracy activist Zhao Jing, who goes by Michael Anti outside China.

128   We are also living in a golden age of periodic publishing on China, where
       knowledgeable insiders share their insights daily or weekly.

       For my money, the single greatest resource for understanding China's day-
       to-day political and economic issues is Bill Bishop's Sinocism (sinocism.
       com) newsletter, which appears at a rate of about three a week. Bishop, a
       Beijing resident until a recent move to the U.S., filters a truly incredible
       range of sources in both English and Chinese, selecting and ordering a small
       list of articles with a paragraph of his informed commentary. He is knowl-
       edgeable, urbane, and cynical (hence the newsletter's title).

       Kaiser Kuo is another source of contemporary observations about China.
       Kuo is the head of international communications for China's search giant
       Baidu. Kuo anchors a weekly podcast, *Sinica* (popupchinese.com/lessons/
       sinica), that brings in guests to discuss contemporary, often media-related,
       events and trends in China. Kuo is also an amazingly prolific and well-
       regarded essayist on Chinese topics on the question-and-answer site Quora
       (quora.com/Kaiser-Kuo).

       88 Bar (88-bar.com) is run by a cadre of artists and social scientists includ-
       ing An Xiao Mina, who studies Chinese popular culture, and Tricia Wang,
       an ethnographer of Chinese technology use. 88 Bar publishes on an erratic
       schedule, but when it does, it is usually worth reading, and their archives
       are quite interesting.

       The Asia Society's "China File" (chinafile.com) is a great source of discus-
       sion of contemporary China. Their roundtable format, where they bring in
       many knowledgeable (and often clashing) perspectives on the questions of
       the day, is particularly worthwhile.

       China Smack (chinasmack.com) translates popular and trending articles
       from Chinese social media. Global Voices (globalvoicesonline.org/-/world/
       east-asia/china/) also translates posts from Chinese bloggers, but
       where China Smack tends toward the popular, Global Voices tends toward
       the political.

       Finally, if you are interested in ongoing news about the Chinese media
       environment, China Internet Watch (chinainternetwatch.com) covers
       mainstream commercial observations, while China Digital Times (china-
       digitaltimes.net) is an indispensible source for tracking the party's
       evolving positions and practices on online censorship and propaganda.

www.indexmundi.com/facts/indi
cators/IT.CEL.SETS.P2/

10   the first new invention added:
"The anthropology of mobile
phones," by Jan Chipchase, TED
Talks, March 2007. http://www.ted.
com/talks/jan_chipchase_on_our_
mobile_phones

**11   fastest-spreading piece of
consumer hardware ever:** "There
are now more gadgets on Earth than
people," by Eric Mack, CNET, Oct. 6,
2014. http://www.cnet.com/news/
there-are-now-more-gadgets-on-
earth-than-people/

**11   Fishermen in Kenya use
phones:** "Fish traders land bigger
returns with market tracking system,"
by Dalton Nyabundi, *Business
Daily Africa,* Jan. 1, 2014. http://www.
businessdailyafrica.com/Fish-
traders-land-bigger-returns-with-
market-tracking-system/-/
1248928/2131390/-/agyo6i/-/index.
html

11   4.5 billion last year: "Number of
mobile phone users worldwide from
2012 to 2018 (in billions)," Statista.
http://www.statista.com/statis-
tics/274774/forecast-of-mobile-
phone-users-worldwide/

**11-12   66 percent in 2014 ... 58
percent ... 25 percent adoption:**
"Mobile cellular subscriptions (per
100 people)," Index Mundi. http://

**14   beating Samsung as the number
one phone vendor in the largest
market in the world in 2014:** "The
China Smartphone Market Picks Up
Slightly in 2014Q4, IDC Reports,"
International Data Corporation, Feb.
17, 2015. http://www.idc.com/get
doc.jsp?containerId=prHK25437515

**14   third largest ecommerce firm
there:** "Xiaomi CEO: Don't call us
China's Apple," Reuters, August
15, 2013. http://www.reuters.com/
video/2013/08/15/xiaomi-ceo-dont-
call-us-chinas-apple?videoId=2490
09264&videoChannel=5

**15   buying Mi.com last year for
$3.6 million:** "Xiaomi widens for-
eign horizons," *China Daily,* April 23,
2014. http://www.chinadaily.com.
cn/business/tech/2014-04/23/con
tent_17458388.htm

**15   the largest online sales day in
any country in the world:** "China's
One-Day Shopping Spree Sets
Record in Online Sales," by Shanshan
Wang and Eric Pfanner, *New York
Times,* Nov. 11, 2013. http://www.
nytimes.com/2013/11/12/business/
international/online-shopping-
marathon-zooms-off-the-blocks-
in-china.html

**15   nearly 1.2 million were
Xiaomis:** "Xiaomi sold nearly 1.2
million phones during China's
24-hour sales bonanza," by Steven

Millward, Tech in Asia, Nov. 13, 2014. https://www.techinasia. com/xiaomi-sold-over-1-million-phones-during-china-singles-day-sales/

15   five of every eight Android phones activated in China were Xiaomis: "Xiaomi has 5 out of the Top 8 Most Activated Android Smartphones," by Alexander Maxham, Android Headlines, Feb. 12, 2015. http://www.androidhead-lines.com/2015/02/xiaomi-5-top-8-activated-android-smartphones. html

15   Xiaomi raised $41 million: "Chinese smartphone maker Xiaomi confirms new $216 million round of funding," by Jon Russell, The Next Web, June 26, 2012. http:// thenextweb.com/asia/2012/06/26/ chinese-smartphone-maker-xiaomi-confirms-new-216-million-round-of-funding/

15   most valuable startup ever: "Xiaomi Becomes World's Most Valuable Tech Startup," by Juro Osawa, Gillian Wong, and Rick Carew, *Wall Street Journal,* Dec. 29, 2014. http:// www.wsj.com/articles/xiaomi-becomes-worlds-most-valuable-tech-startup-1419843430

17-18   Big Mac index: "The Big Mac index," *The Economist,* Jan. 22, 2015. http://www.economist.com/content/big-mac-index

21   Studies of censored topics: 131
"China's Censorship 2.0: How com panies censor bloggers," by Rebecca MacKinnon, *First Monday,* Vol. 14, No. 2, Feb. 2, 2009. http:// firstmonday.org/article/ view/2378/2089

23   information would behave more like money: "Google CEO: China's Internet censorship will fail in time," by Michael Kan, IDG News Service, Nov. 4, 2010. http://www. computerworld.com/article/ 2513905/internet/google-ceo-china-s-internet-censorship-will-fail-in-time.html "China's censorship can never defeat the internet," by Ai Weiwei, the *Guardian,* April 15, 2012. http:// www.theguardian.com/comment isfree/libertycentral/2012/apr/16/ china-censorship-internet-freedom

23   spend more on internal security than on their military: "China hikes defense budget, to spend more on internal security," by Ben Blanchard and John Ruwitch, Reuters, March 5, 2013. http://www. reuters.com/article/2013/03/05/ us-china-parliament-defence-idUSBRE92403620130305

25-26   recruiting members of the Communist Youth League ... Beijing is expected to recruit ... Sun Yat-sen University: "Wanted: Ten million Chinese students to 'civilize' the Internet," by Xu Yangjingjing and Simon Denyer, *Washington Post,* April 10,

132 2015. http://www.washington
post.com/blogs/worldviews/
wp/2015/04/10/wanted-ten-
million-chinese-students-to-
civilize-the-internet/

26 **recruit one-fifth of the
Communist Youth League:**
"Leaked Emails Reveal Details of
China's Online 'Youth Civilization
Volunteers,' by Patrick Wong, Global
Voices, May 25, 2015. http://
globalvoicesonline.org/2015/05/25/
leaked-mails-reveal-details-on-
chinas-online-youth-civilization-
volunteers/

26 **"the tearing apart of social
consensus":** "Army Newspaper: We
Can Absolutely Not Allow the
Internet Become a Lost Territory of
People's Minds," China Copyright
and Media, May 13, 2015. https://
chinacopyrightandmedia.wordpress.
com/2015/05/13/army-newspaper-
we-can-absolutely-not-allow-the-
internet-become-a-list-territory-
of-peoples-minds/
Originally published in Chinese in
the *People's Liberation Army Daily,*
May 12, 2015. http://news.mod.gov.
cn/headlines/2015-05/12/
content_4584573.htm

27 **long-debated national security
law:** "National Security Law," Section
4, Article 59, China Law Translate.
http://chinalawtranslate.com/
en/2015nsl/

33 **"fever" fans... and "flood"
fans:** "Fan-centric social media: The

Xiaomi phenomenon in China," by
Chao-Ching Shih, Tom M.Y. Lin, and
Pin Luarn, *Business Horizons*, Vol. 57,
Issue 3, May—June 2014, pp. 349—
358. http://www.sciencedirect.com/
science/article/pii/
S0007681313002140

37 **continue to prefer simple
"monoblock" and flip phones:**
"Nokia new models out," by Zheng
Lifei, *China Daily,* June 26, 2009.
http://www.chinadaily.com.cn/
bizchina/2009-06/26/content_
8325265.htm

45 **widely circulated report:**
"Testing the Xiaomi RedMi 1S,"
F-Secure, August 7, 2014. https://
www.f-secure.com/weblog/
archives/00002731.html

46 **set the defaults for data
sharing and messaging to Off:**
"MIUI Cloud Messaging & Privacy,"
by Hugo Barra, Google+, August 10,
2014. https://plus.google.com/+
HugoBarra/posts/bkJTXzyXXmj

46 **Indian Air Force asked that its
personnel not buy Xiaomi phones:**
"Updated: Xiaomi spying on users
and forwarding personal information
to China says IAF," by Aparajita
Saxena, MediaNama, Oct. 22,
2014. http://www.medianama.
com/2014/10/223-xiaomi-spying-
on-users/

46 **personal data being sent to
servers in Beijing:** "Xiaomi under-
investigation for sending user info

back to China," by Liu Jiayi, ZDNet, Sept. 11, 2014.http://www.zdnet. com/article/xiaomi-under- investigation-for-sending-user- info-back-to-china/

46   **selling users' numbers to telemarketers:** "Singapore Investigating Data Complaint Against Xiaomi," by Newley Purnell, *Wall Street Journal,* August 15, 2014. http://blogs.wsj.com/digits/2014 /08/15/singapore-investigating -data-complaint-against-xiaomi/

47   **"It is theft and it is lazy":** "Apple's Jony Ive Is Not Flattered By Xiaomi," by Kyle Russell, Tech Crunch, Oct. 9, 2014. http://tech crunch.com/2014/10/09/apples- jony-ive-is-not-flattered-by- xiaomi/

56   **suicide of four children in Guizhou province:** "Chinese police 'find suicide note' in case of 'left behind' children deaths," by Tom Phillips, the *Guardian,* June 14, 2015. http://www.theguardian.com/ world/2015/jun/14/chinese-police- investigating-deaths-of-left- behind-children-find-suicide-note

68   **The company had effectively no profit in 2011:** "China's Xiaomi booked $56 million profit in 2013," by Gerry Shih, Reuters, Dec. 16, 2014. http://www.reuters.com/arti- cle/2014/12/16/us-xiaomi-finan- cials-idUSKBN0JT07Y20141216

70   **40,000 of their cheap RedMi**   133 **1S phones in *four seconds*:** "Xiaomi sells 40,000 Redmi 1S phones in 4 seconds in India," by Aloysius Low, CNET, Sept. 2, 2014. http://www. cnet.com/uk/news/xiaomi-sells- out-40000-redmi-1s-in-4- seconds-in-india/

79   **"Xiaomi's mission is to change the world's view of Chinese prod- ucts":** "Xiaomi, China's New Phone Giant, Takes Aim at World," by Eva Dou, *Wall Street Journal,* June 7, 2015. http://www.wsj.com/articles/ xiaomi-chinas-new-phone-giant- takes-aim-at-world-1433731461

91   **Much more: 6.6 billion tons to 4.5:** "How China used more cement in 3 years than the U.S. did in the entire 20th Century," by Ana Swanson, *Washington Post,* March 24, 2015. http://www. washingtonpost.com/blogs/ wonkblog/wp/2015/03/24/ how-china-used-more-cement- in-3-years-than-the-u-s-did-in- the-entire-20th-century/

99   **"deeply reflects the Chinese people's dream today":** "Chinese Dreams," by Geremie R. Barmé, The China Story. http://www.thechinas- tory.org/yearbooks/yearbook-2013/ forum-dreams-and-power/chinese- dreams-zhongguo-meng-%E4% B8%AD%E5%9B%BD%E6% A2%A6/

134   104   **China is remarkably, overwhelmingly homogenous:** "The
World Factbook: China," Central
Intelligence Agency. https://www.
cia.gov/library/publications/
the-world-factbook/geos/ch.html

107   **recent crackdown on VPNs:**
"China intensifies VPN services
crackdown," by Charles Clover,
Financial Times, Jan. 23, 2015. http:
//www.ft.com/intl/cms/
s/0/46ad9e26-a2b9-11e4-9630-
00144feab7de.html

116   **Samsung's market share:**
"Smartphone Vendor Market
Share, Q1 2015," International Data
Corporation. http://www.idc.com/
prodserv/smartphone-market-
share.jsp